Quantifying Life:
A Symbiosis of Computation, Mathematics, and Biology

Dmitry A. Kondrashov

The University of Chicago Press
Chicago and London

The University of Chicago Press, Chicago 60637
The University of Chicago Press, Ltd., London
© 2016 by Dmitry A. Kondrashov
All rights reserved. Published 2016.
Printed in the United States of America

25 24 23 22 21 20 19 18 17 16   1 2 3 4 5

ISBN-13: 978-0-226-37162-7   (cloth)
ISBN-13: 978-0-226-37176-4   (paper)
ISBN-13: 978-0-226-37193-1   (e-book)
DOI: 10.7208/chicago/9780226371931.001.0001

Library of Congress Control Number: 2016939464

♾ This paper meets the requirements of ANSI/NISO Z39.48-1992 (Permanence of Paper).

To my teachers, for challenging me when I was complacent
and for helping me rise to the challenge

Science begins with counting. To understand a phenomenon, a scientist must first describe it; to describe it objectively, he must first measure it.

—Siddhartha Mukherjee, *The Emperor of All Maladies*

Contents

Preface   xiii

0 Introduction   1
    0.1 What is mathematical modeling? 1
    0.2 Purpose of this book 2
    0.3 Organization of the book 4

I Describing single variables   7

1 Arithmetic and variables: The lifeblood of modeling   9
    1.1 Blood circulation and mathematical modeling 11
    1.2 Parameters and variables in models 15
        1.2.1 discrete state variables: genetics 17
        1.2.2 continuous state variables: concentration . . 18
    1.3 First steps in R programming 19
        1.3.1 numbers and arithmetic operations 20
        1.3.2 variable assignment 23
    1.4 Computational projects 26

2 Functions and their graphs   29
    2.1 Dimensions of quantities 30
    2.2 Functions and their graphs 33
        2.2.1 linear and exponential functions 34
        2.2.2 rational and logistic functions 37
    2.3 Scripts, functions, and plotting in R 41

		2.3.1	writing scripts and calling functions	41
		2.3.2	vector variables	42
		2.3.3	arithmetic with vector variables	45
		2.3.4	plotting graphs	47
	2.4	Rates of biochemical reactions		51
	2.5	Computational projects		53

3 Describing data sets 57
3.1 Mutations and their rates 57
3.2 Describing data sets 60
3.2.1 central value of a data set 60
3.2.2 spread of a data set 63
3.2.3 graphical representation of data sets 65
3.3 Working with data in R 67
3.4 Computational projects 70
3.4.1 data description 70
3.4.2 plotting data and fitting by eye 71

4 Random variables and distributions 75
4.1 Probability distributions 76
4.1.1 axioms of probability 76
4.1.2 random variables 81
4.1.3 expectation (mean) of random variables . . . 82
4.1.4 variance of random variables 84
4.2 Examples of distributions 86
4.2.1 uniform distribution 86
4.2.2 binomial distribution 87
4.3 Testing for mutants 93
4.4 Random numbers and iteration in R 94
4.4.1 random numbers 94
4.4.2 for loops 97
4.5 Computational projects 100
4.5.1 uniform distribution 100
4.5.2 binomial distribution 101

CONTENTS

5 Estimation from a random sample — 103
- 5.1 Law of Large Numbers — 104
  - 5.1.1 sample mean — 104
  - 5.1.2 sample size and standard error — 106
- 5.2 Central Limit Theorem — 107
  - 5.2.1 normal distribution — 107
  - 5.2.2 confidence intervals — 110
- 5.3 Relative risk — 114
- 5.4 Sampling in R — 116
  - 5.4.1 simulated sampling — 116
  - 5.4.2 computing confidence intervals — 118
- 5.5 Computational projects — 121

II Relationship between two variables — 123

6 Independence of random variables — 125
- 6.1 Categorical data sets with two variables — 126
- 6.2 Mathematics of independence — 127
  - 6.2.1 conditional probability and information — 127
  - 6.2.2 independence of events — 132
  - 6.2.3 calculation of expected frequencies — 133
- 6.3 Testing for independence — 137
  - 6.3.1 hypothesis testing — 137
  - 6.3.2 rejecting the null hypothesis — 140
  - 6.3.3 the chi-squared statistic — 142
- 6.4 Hypothesis testing in R — 145
- 6.5 Independence in data sets — 147
  - 6.5.1 maternal age and Down syndrome — 147
  - 6.5.2 stop-and-frisk and race — 148
- 6.6 Computational projects — 150
  - 6.6.1 thumb-on-top preference and sex — 150
  - 6.6.2 relationship between species and habitat — 151
  - 6.6.3 independence testing of simulated data — 152

7	**Bayes' amazing formula**	**155**
	7.1 Prior knowledge	156
	7.2 Bayes' formula	157
	7.2.1 positive and negative predictive values	159
	7.3 Applications of Bayesian thinking	162
	7.3.1 when too much testing is bad	162
	7.3.2 reliability of scientific studies	164
	7.4 Random simulations	167
	7.5 Computational project	172
8	**Linear regression and correlation**	**177**
	8.1 Linear relationship between two variables	178
	8.2 Linear least-squares fitting	178
	8.2.1 sum of squared errors	178
	8.2.2 best-fit slope and intercept	182
	8.2.3 correlation and goodness of fit	184
	8.3 Linear regression using R	188
	8.4 Regression to the mean	191
	8.5 Computational projects	193
	8.5.1 parental age and new mutations	194
	8.5.2 heart rates on two different days	195
9	**Nonlinear data fitting**	**197**
	9.1 Nonlinear relationships between variables	197
	9.2 Fitting using log transforms	199
	9.3 Generalized linear fitting in R	201
	9.3.1 logarithmic transforms	201
	9.3.2 polynomial regression	202
	9.4 Allometry and power law scaling	205
	9.5 Computational projects	208
III	**Chains of random variables**	**211**
10	**Markov models with discrete states**	**213**
	10.1 Building Markov models	214

	10.2	Markov property	216
		10.2.1 transition matrices	219
		10.2.2 probability of a string of states	220
	10.3	Simulation of Markov models	224
	10.4	Markov models of medical treatment	226
	10.5	Computational projects	229
		10.5.1 state strings for a two-state model	229
		10.5.2 state strings for a three-state model	230

11 Probability distributions of Markov chains — 233

	11.1	Distributions evolve over time	234
		11.1.1 Markov chains	235
		11.1.2 matrix multiplication	237
		11.1.3 propagation of probability vectors	241
	11.2	Matrix multiplication in R	244
	11.3	Mutations in evolution	247
	11.4	Computational projects	249
		11.4.1 probability vectors of a two-state model	250
		11.4.2 probability vectors of a three-state model	251

12 Stationary distributions of Markov chains — 253

	12.1	The origins of Markov chains: A feud and a poem	254
	12.2	Stationary distributions	256
		12.2.1 definition of stationary distribution	256
		12.2.2 condition for unique stationary distribution	259
	12.3	Multiple random simulations in R	264
	12.4	Bioinformatics and Markov models	265
	12.5	Computational projects	270
		12.5.1 multiple two-state model simulations	270
		12.5.2 multiple three-state model simulations	272

13 Dynamics of Markov models — 275

	13.1	Phylogenetic trees	276
	13.2	Eigenvalues of Markov models	278
		13.2.1 basic linear algebra terminology	278
		13.2.2 calculation of eigenvalues on paper	281
		13.2.3 calculation of eigenvectors on paper	283
		13.2.4 rate of convergence	285

13.3 Matrix diagonalization in R 288
13.4 Molecular evolution 293
    13.4.1 Jukes-Cantor model 293
    13.4.2 time since divergence 296
    13.4.3 calculation of phylogenetic distance 297
    13.4.4 divergence of human and chimp genomes . . 298
13.5 Computational projects 301
    13.5.1 eigenvalues of a two-state model 301
    13.5.2 eigenvalues of a three-state model 302
    13.5.3 analysis of the Jukes-Cantor model 303

IV Variables that change with time   305

14 Linear difference equations   307
14.1 Discrete-time population models 308
    14.1.1 static population 308
    14.1.2 exponential population growth 309
    14.1.3 example with birth and death 309
    14.1.4 dimensions of birth and death rates 311
14.2 Solutions of linear difference models 312
    14.2.1 simple linear models 312
    14.2.2 models with a constant term 313
14.3 Population growth and decline 314
14.4 Numerical solutions in R 317
    14.4.1 functions in R 317
    14.4.2 solving difference equations 319
14.5 Computational project 322

15 Linear ordinary differential equations   325
15.1 From discrete time to smooth change 326
    15.1.1 bacteria that divide at arbitrary times 326
    15.1.2 growth proportional to population size 329
    15.1.3 chemical kinetics 330
15.2 Solutions of ordinary differential equations 331
    15.2.1 separate-and-integrate method 332
    15.2.2 solution of inhomogeneous ODEs 335
    15.2.3 Forward Euler method 338

15.3	Numerical solutions of ODEs	341
	15.3.1 implementation of Forward Euler	341
	15.3.2 error in Forward Euler solutions	343
15.4	Applications of linear ODE models	347
	15.4.1 model of pharmacokinetics	347
	15.4.2 Cole's membrane potential model	351
15.5	Computational projects	355
	15.5.1 error and time step	355
	15.5.2 pharmacokinetics model	356

16 Graphical analysis of ordinary differential equations — 359

16.1	ODEs with nonlinear terms	360
16.2	Qualitative analysis of ODEs	361
	16.2.1 graphical analysis of the defining function	362
	16.2.2 fixed points and stability	364
16.3	Modeling infectious disease spread	372
16.4	Computational projects	377
	16.4.1 logistic population growth model	378
	16.4.2 SIS epidemic model	379

17 Chaos and bifurcations in difference equations — 383

17.1	Logistic model in discrete time	384
17.2	Qualitative analysis of difference	385
	17.2.1 fixed points or equilibria	385
	17.2.2 stability of fixed points	386
17.3	Graphical analysis	389
	17.3.1 graphical analysis using R	389
	17.3.2 cobweb plots	390
17.4	Discrete-time logistic model and chaos	395
17.5	Computational projects	402
	17.5.1 graphical stability analysis	402
	17.5.2 investigation of chaotic dynamics	403

Bibliography — 405

Index — 411

Preface

I wrote this textbook for a course that I developed for first-year college students intending to study biology at the University of Chicago as a part of curriculum reform initiated by the Master of Biological Sciences Collegiate Division José Quintáns. Biology is changing at a break-neck pace and is becoming ever more reliant on complex computational tools and quantitative models, while the curriculum for biology majors has been largely stuck in the pre-genomics era. This is especially true of the quantitative component, which is traditionally limited to (at most) a year of calculus and a statistics course. This is not enough to even begin to follow the ideas and methods of current research, which heavily relies on stochastic models, differential equations, and computational algorithms.

In the last few years, these challenges have been recognized by the biology community and have resulted in a boom in bio-calculus textbooks and materials. While combining the concepts of calculus (derivatives, integrals, series) with biological applications sounds good, in my view it still doesn't supply the critical missing components mentioned above. In fact, I started out teaching this course using a bio-calculus approach but realized that this severely restricts the range of biological applications that could be discussed. I also found that the traditional mathematics pedagogy of working out problems on paper is not sufficient for addressing many questions of interest. Instead, I realized that computation is the most useful, as well as the most challenging, skill needed for a quantitative understanding of modern biology.

These hard-won empirical observations led me to develop the course and the textbook as a chimeric beast made up of elements of statistics, probability, Markov models, and differential equations. The thread that holds these pieces together is programming, which enables students to see and experience fairly complex mathematical ideas. Therefore, each chapter in the textbook contains a modeling section, a mathematics section, and computational section, which all reinforce each other and provide different viewpoints on the same concept. Based on my experience, students with no prior programming experience find the first steps to be tough going, but by the end of course they often achieve a level of comfort with basic coding that they find very valuable in their research and education.

As mentioned above, the person who is most responsible for the existence of this book is José Quintáns. He had a vision of a biology curriculum that served the needs of students and firmly believed in the necessity of greater quantitative literacy in this field. He allowed me to develop my courses the way I saw fit, while providing support and guidance when needed, and encouraged me to start writing up my notes as a book. I am greatly indebted to him, so I thank him and wish him a happy and active retirement.

I also want to thank my colleagues in Biological Sciences Collegiate Division for providing an excellent work environment, and in particular Esmael Haddadian and Elizabeth Kovar, who helped me teach this course in earlier iterations. Donald Frederick, a former student who later worked as a TA and then served as computer lab director for the course, helped me greatly with finding the right approach for the computational assignments. I also thank all of the TAs over the years, whose contributions and feedback have helped shape the course: Colin Pesyna, Chris Bun, Lamont Samuels, Pooya Hatami, Denis Pankratov, Sandra Fernandez, Negar Mirsattari, Mladen Rasic, Kristen Voorhies, Tim Armstrong, Erik Bodzsar, Alexander Ostapenko, Tasneem Amina, Jason McCreery, Andrea Garofalo, Davis Bennet, Aleks Penev, Kirstie Wade, Sven Auhagen, Sneha Popley, Sobha Naderi, Matthew Battifarano, Brad Cohn, Adam Filipowicz, Hunter Chase, Erika Dunn-Weiss, Charles Frye, and Lynda Lin-Shiu.

I thank the colleagues and friends who read the manuscript at various stages of its evolution and helped improve it with their suggestions: Will Trimble, Sarah Hews, Hannah Callender, Kayla

PREFACE

Lewis, Jeff Edmunds, Daniel Coombs, Natalia Toporikova, John Jungck, Carl Bergstrom, and Timofey Kondrashov. I also thank my wife, Arielle Hirschfeld, for bringing in her medical perspective, which is reflected in many applications in the text. I thank the editors and staff at the University of Chicago Press, particularly Christie Henry, Jenni Fry, Nick Lilly, and Logan Smith, for patiently shepherding the book from a manuscript and for handling the promotion and production details. I'm also greatly indebted to Mikala Guyton and Cyd Westmoreland at Westchester Publishing Services for supervising the editorial process and for copyediting the manuscript with great care, respectively.

This book was written and typeset using a number of open-source tools, so I would like to thank the developers and the communities that maintain and improve these phenomenal resources and keep them available to all. I use the R language https://www.r-project.org/ for programming and generating many of the figures in the text and R Studio https://www.rstudio.com/ as the user-friendly interface for working with R. The manuscript was typeset using LaTeX https://www.latex-project.org/ with the MacTeX distribution https://tug.org/mactex/ and the TeXShop front end http://pages.uoregon.edu/koch/texshop/obtaining.html. The combination of R and LaTeX was done using the knitr package http://yihui.name/knitr/ and many of the figures were generated using the TikZ graphics package http://www.texample.net/tikz/. I also thank the biomathematics community and specifically acknowledge the new QUBES hub https://qubeshub.org/ for providing a repository for teaching resources and for connecting colleagues teaching at the interface of mathematics, computation, and biology.

This book would not exist without the teachers whom I've been privileged to learn from, starting from middle school (School 2 in Pushchino, Russia), Memorial High School in Madison, Wisconsin, Simon's Rock College of Bard, and the Graduate Program in Applied Mathematics at the University of Arizona. The teachers and mentors to whom I am most indebted are Yelena Gavrilovna Kuznetsova, Victor Levine, William Dunbar, Eileen Handelman, Joseph Watkins, Hermann Flaschka, Joceline Lega, and William

Montfort. They all helped me learn and grow, and continue to inspire me today.

My family has encouraged my curiosity and love of science since I first became conscious. For this I thank my parents, Alexey and Natalia, as well as my grandparents, Maria Nikolayevna Kondrashova and Simon Elievich Shnol, who are both scientists themselves, and Nina Ivanovna Vasilieva and Yuri Semyonovich Sedov, who also encouraged and supported me. I am lucky to have the best brothers anyone could have: Fedya, Vasya, Misha, and Tim. I am profoundly grateful to my wife, Arielle Hirschfeld, who had to put up with years of my work on this manuscript and has not only supported and encouraged me in this endeavor but also made substantial contributions, as I mentioned above. Finally, my son Ellis has also helped with his interest in my work and his example of working on his own writing projects, often much more efficiently than his father.

Chapter 0

Introduction

> *"What is a man," said Athos, "who has no landscape? Nothing but mirrors and tides."*
> —Anne Michaels, *Fugitive Pieces*

0.1 What is mathematical modeling?

A *mathematical model* is a representation of some real object or phenomenon in terms of quantities (numbers). The goal of modeling is to create a description of the object in question that may be used to pose and answer questions about it without doing hard experimental work. A good analogy for a mathematical model is a map of a geographic area: a map cannot record all the complexity of the actual piece of land, because the map would need to be size of the piece of land, and then it wouldn't be very useful! Maps, and mathematical models, need to sacrifice the details and provide a bird's-eye view of reality to guide the traveler or the scientist. The representation of reality in the model must be simple enough to be useful, yet complex enough to capture the essential features of what it is trying to represent.

Since the time of Newton, physicists have been very successful at using mathematics to describe the behavior of matter of all sizes, ranging from subatomic particles to galaxies. However, mathematical modeling is a new arrow in a biologist's quiver. Many biologists would argue that living systems are much more complex than either atoms or galaxies, since even a single cell is made up of a mind-boggling number of highly dynamic, interacting entities. This complexity presents a great challenge and fascinating new questions.

New advances in experimental biology are producing data that make quantitative methods indispensable for biology. The advent of *genetic sequencing* in the 1970s and 1980s has allowed us to determine the genomes of different species, and in the past few years next-generation sequencing has reduced sequencing costs for an individual human genome to a few thousand dollars. The resulting deluge of quantitative data has answered many outstanding questions and has also led to entirely new ones. We now understand that knowledge of genomic sequences is not enough for understanding how living things work, so the burgeoning field of *systems biology* investigates the interactions among genes, proteins, or other entities. The central problem is to understand how a network of interactions among individual molecules can lead to large-scale results, such as the development of a fertilized egg into a complex organism. The human mind is not suited for making correct intuitive judgements about networks comprised of thousands of actors. Addressing questions of this complexity requires quantitative modeling.

0.2 Purpose of this book

This textbook is intended for a college-level course for biology and pre-medicine majors, or for more established scientists interested in learning the applications of mathematical methods to biology. The book brings together concepts found in mathematics, computer science, and statistics courses to provide the student a collection of skills that are commonly used in biological research. The book has two overarching goals. The first is to explain the quantitative language that often is a formidable barrier to understanding and critically evaluating research results in biological and medical sciences. The second is to teach students computational skills that they can use in their future research endeavors. The main premise of this approach is that computation is critical for understanding abstract mathematical ideas.

0.2. PURPOSE OF THIS BOOK

These goals are distinct from those of traditional mathematics courses that emphasize rigor and abstraction. I strongly believe that understanding mathematical concepts is not contingent on being able to prove all of the relevant theorems. Instead, premature focus on abstraction obscures the ideas for most students; it is putting the theoretical cart before the experiential horse. I find that students can grasp deep concepts when they are allowed to experience them tangibly as numbers or pictures, and those with an abstract mindset can generalize and add rigor later. As I demonstrate in part 3 of the book, Markov chains can be explained without relying on the machinery of measure theory and stochastic processes, which require graduate-level mathematical skills. The idea of a system randomly hopping between a few discrete states is far more accessible than sigma algebras and martingales. Of course, some abstraction is necessary when presenting mathematical ideas, and I provide correct definitions of terms and supply derivations when I find them to be illuminating. But I avoid rigorous proofs and always favor understanding over mathematical precision.

The book is structured to facilitate learning computational skills. Over the course of the text, students accumulate programming experience, progressing from assigning values to variables in Chapter 1 to solving nonlinear Ordinary differential equations (ODEs) numerically by the end of the book. Learning to program for the first time is a challenging task, and I facilitate it by providing sample scripts for students to copy and modify to perform the requisite calculations. Programming requires careful, methodical thinking, which facilitates deeper understanding of the models being simulated. In my experience teaching this course, students consistently report that learning basic scientific programming is a rewarding experience, which opens doors for them in future research and learning.

It is of course impossible to span the breadth of mathematics and computation used for modeling biological scenarios. This did not stop me from trying. The book is broad but selective, sticking to a few key concepts and examples that should provide enough of a basis for a student to explore a topic in more depth later on. For instance, I do not go through the usual menagerie of probability distributions in Chapter 4 but only analyze the uniform and the binomial distributions. If one understands the concepts of distributions and their means and variances, it is not difficult to read up on the geometric or gamma distribution if one encounters it. Still, I omitted numerous topics and entire fields, some because they require greater mathematical sophistication, and others because they are too difficult for beginning programmers (e.g., sequence alignment and optimization algorithms). I hope that you do not end your quantitative journey with this book!

I take an even more selective approach to the biological topics presented in every chapter. The book is not intended to teach biology, but I do introduce biological questions I find interesting, refer to current research papers, and provide discussion questions for you to wrestle with. This requires a basic explanation of terms and ideas, so most chapters contain a broad summary of a biological field, such as measuring mutation rates, epidemiology modeling, hidden Markov models for gene structure, and limitations of medical testing. I hope the experts in these fields forgive my omitting the interesting details that they spend their lives investigating, and trust that I managed to get the basic ideas across without gross distortion.

0.3 Organization of the book

Each chapter in the textbook is centered around a mathematical concept, along with models, biological applications, and programming. This multipronged approach provides a diverse set of teaching tools: motivational questions from biology can be formalized using mathematical terms, solved for simple cases on the board, and then demonstrated in more complex manifestations using the programming language R. Each chapter contains enough material for a week of learning and includes various assignments. The mathematics sections contain simple practice problems for the corresponding mathematical skills, the programming sections contain either debugging exercises or simple programming assignments, and the biological modeling sections contain discussion questions intended to stimulate students to think about assumptions and limitations

0.3. ORGANIZATION OF THE BOOK

of the models (and they frequently require students to read and digest a research paper). Each chapter ends with multi-question computational projects that walk students through implementing and investigating a computational model for a biological question.

Part 1 of the textbook (Chapters 1–5) starts with elementary mathematical ideas: variables and parameters, basic functions and graphs, and descriptive statistics. These simple concepts pair well with rudimentary programming steps that are introduced concurrently. Despite the conceptual simplicity, the first attempts at writing and executing code are invariably difficult for students, so I find this combination pedagogically sound. More advanced students can treat the first three chapters as review, but those who have never written code before are advised to focus on the programming exercises. Chapters 4 and 5 are less elementary, and students may encounter something new in the realms of probability distributions and estimation through sampling.

Part 2 of the book (Chapters 6–9) concerns relationships between two variables, both categorical and numerical. This is a largely data-driven part of the course, but it also introduces crucial theoretical concepts that are used later, particularly conditional probability and independence. I present the standard chi-squared test for independence and then warn students about misuse of p-values in the chapter on Bayesian thinking. The ideas of linear regression are familiar to most students at this level, but few are acquainted with correlation at a more than perfunctory level. The last chapter of this part delves into nonlinear fitting using logarithmic transformations and its applications.

Part 3 of the book (Chapters 10–13) is an introduction to Markov models divided into four chapters. The story progresses from describing models with transition matrices and flow diagrams to recursive calculation of probability distribution vectors, then to stationary distributions and finally to describing dynamics using eigenvalues and eigenvectors. The level of mathematical sophistication jumps considerably, and so do the computational expectations. Students learn to generate simulated strings of Markov states and then to repeat the simulations to generate entire data sets evolving over time.

Part 4 of the book (Chapters 14–17) addresses one-variable dynamical systems. The first chapter analyzes linear discrete-time equations and their solutions; the next one graduates to linear differential equations and their solutions, which build on the discrete-time ideas. We then move to graphical analysis of nonlinear ODEs, and finish with a look at the crazy behavior and chaos in nonlinear discrete-time models.

A one-semester (or one-quarter) course based on this book can be designed in several ways. The first two parts of the book provide the necessary foundation for the next two, both mathematically and in programming skills, but parts 3 and 4 are essentially independent. One could teach a reasonable course based on either parts 1, 2, and 3, or parts 1, 2, and 4. Another option is to omit the last chapter of each part (Chapters 5, 9, 13, and 17), because they contain more advanced topics than the rest and are designed to be skipped without any detriment to the flow of ideas. I should note that with the exception of part 4 (actually only the last three chapters), none of the rest use any concepts from calculus, so one could design a course for students with shaky or nonexistent knowledge of calculus. For an audience with greater mathematical maturity, one could power through part 1 in 2–3 weeks and be able to go through most of the textbook in a semester.

A course based on this textbook can be tailored to fit the quantitative needs of a biological sciences curriculum. At the University of Chicago, the course I teach has replaced the last quarter of calculus as a first-year requirement for biology majors. This material could be used for a course without a calculus prerequisite that a student takes before more rigorous statistics, mathematics, or computer science courses. It may also be taught as an upper-level elective course for students with greater maturity who may be ready to tackle the chapters on eigenvalues and differential equations. My hope is that it may also prove useful for graduate students or established scientists who need an elementary but comprehensive introduction to the concepts they encounter in the literature or that they can use in their own research. Whatever path you traveled to get here, I wish you a fruitful journey through biomathematics and computation!

Part I
Describing single variables

Chapter 1

Arithmetic and variables: The lifeblood of modeling

You can add up the parts, but you won't have the sum;
You can strike up the march, there is no drum.
Every heart, every heart to love will come
But like a refugee.

—Leonard Cohen, *Anthem*

Mathematical modeling begins with a set of *assumptions*. In fact, one may say that a mathematical model is a bunch of assumptions translated into mathematics. These assumptions may be more or less reasonable, and they may come from different sources. For instance, many physical models are so well established that we refer to them as laws; we are pretty sure they apply to molecules, cells, and organisms as well as to inanimate objects. Thus at times we may use physical laws as the foundation on which to build models of biological entities; these are often known as *first-principles* (theory-based) models. At other times we may have experimental evidence that suggests a certain kind of relationship between quantities—perhaps we find that the amount of administered drug and the time until the drug is completely removed from the bloodstream are proportional to each other. This observation can be turned into an *empirical* (experiment-based) model. Yet another type of model assumption is not based on either theory or experiment, but simply on convenience: for example, we may assume that the mutation rates at two different loci are independent and see what the implications are. These are sometimes called *toy* or *cartoon* models (Jungck, Gaff, and Weisstein 2010).

This leads to the question: how do you decide whether a model is good? It is surprisingly difficult to give a straightforward answer to this question. Of course, one major goal of a model is to capture some essential features of reality, so in most biological modeling studies you will see a comparison between experimental results and predictions of the model. But it is not enough for a model to be faithful to experimental data! Think of a simple example: suppose your experiment produced 5 data points as a function of time; it is possible to find a polynomial (of fourth degree) that passes exactly through all 5 points by specifying the coefficients of its 5 terms. This is called *data fitting*, and it has a large role to play in the mathematical modeling of biology. However, I think you will agree that in this case we have learned very little: we just substituted 5 values in the data set with 5 values of the coefficients of the mathematical model. To heighten the absurdity, imagine a data set of 1001 points that you have modeled using a 1000-degree polynomial. This is an example of *overfitting*, or making the model agree with the data by making the model overly complex.

Substituting a complicated model for a complicated real situation does not help understand the reality. One necessary ingredient of a useful model is *simplicity of assumptions*. Simplicity in modeling has at least two virtues: simple models can be grasped by our limited minds, and simple assumptions can be tested against evidence. A simple model that fails to reproduce experimental data can be more informative than a complex model that fits the data perfectly. If a simple model fails, you have learned that you are missing something in your assumptions; but a complex model can be right for the wrong reasons, like erroneous assumptions canceling each other, or it may contain needless assumptions. This is why the ability to build good models is a difficult skill that balances simplicity of assumptions against fidelity to empirical data (Cohen 2004). In this chapter you will learn how to do the following:

1. distinguish variables and parameters in models,
2. describe the state space of a model,
3. perform arithmetic operations in R, and
4. assign variables in R.

1.1 Blood circulation and mathematical modeling

Galen was one of the great physicians of antiquity. He studied how the body works by performing experiments on humans and animals. Among other things, he was famous for a careful study of the heart and how blood traveled through the body. Galen observed that there were different types of blood: arterial blood that flowed out of the heart, which was bright red, and venous blood that flowed in the opposite direction, which was a darker color. This naturally led to questions: what is the difference between venous and arterial blood? Where does each one come from and where does it go?

You, a reader of the twenty-first century, likely already know the answer: blood *circulates* through the body, bringing oxygen and nutrients to the tissues through the arteries, and returns back through the veins carrying carbon dioxide and waste products, as shown in Figure 1.1. Arterial blood contains a lot of oxygen, while venous blood carries more carbon dioxide, but otherwise they are the same fluid. The heart does the physical work of pushing arterial blood out of the heart, to the tissues and organs, as well as pushing venous blood through the second circulatory loop that goes through the lungs, where it picks up oxygen and releases carbon dioxide, becoming arterial blood again. This may seem like a very natural picture to you, but it is far from easy to deduce by simple observation.

Galen came up with a different explanation based on the notion of "humors," or fluids, that was fundamental to the Greek conception of the body. He proposed that the venous and arterial blood were different humors: venous blood, or "natural spirits," was produced by the liver, while arterial blood, or "vital spirits," was

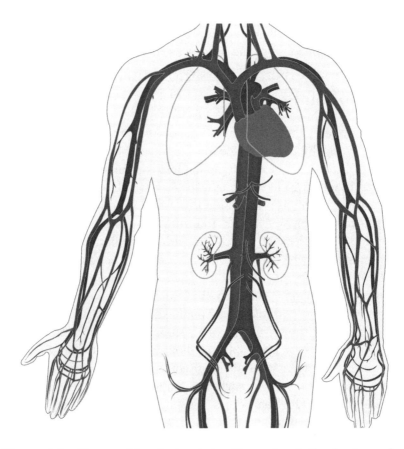

Figure 1.1: Human blood circulates throughout the body and returns to the heart. Veins are shown in blue (dark gray) and arteries in red (light gray). Figure "Circulatory System" by LadyofHats in public domain via Wikimedia Commons.

1.1. BLOOD CIRCULATION

produced by the heart and carried by the arteries, as shown in Figure 1.2. The heart consisted of two halves, and it warmed the blood and pushed both the natural and vital spirits out to the organs; the two spirits could mix through pores in the septum separating its right and left halves. The vital and natural spirits were both consumed by the organs, and they were regenerated by the liver and the heart. The purpose of the lungs was to serve as bellows, cooling the blood after it was heated by the heart.

Is this a good theory of how the heart, lungs, and blood work? Doctors in Europe thought so for more than a thousand years! Galen's textbook on physiology was the standard for medical students through the seventeenth century. The theory seemed to make sense and explain what was observable. Many great scientists and physicians, including Leonardo da Vinci and Avicenna, did not challenge the inaccuracies, such as the porous septum in the heart, even though they could not see the pores themselves. It took both better observations and a quantitative testing of the hypothesis to challenge the orthodoxy.

William Harvey was born in England and studied medicine in Padua under the great physician Hieronymus Fabricius. He became famous and would perform public demonstrations of physiology, using live animals for experiments that would not be legal today. He also studied the heart and the blood vessels, and he measured the volume of the blood that can be contained in the human heart. He was quite accurate in estimating the correct volume, which we now know to be about 70 mL (1.5 oz). What is even more impressive is that he used this quantitative information to test Galen's theory.

Let us assume that all of the blood pumped out by the heart is consumed by the tissues, as Galen proposed; let us further assume that the heart beats at constant rate of 60 beats per minute, with a constant ejection volume of 70 mL. Then over the course of a day, the human body would consume about 70 mL $\times$ 60 (beats per minute) $\times$ 60 (minutes per hour) $\times$ 24 (hours per day), which is more than 6,000 liters of blood! You may quibble over the exact numbers (some hearts beat faster or slower, some hearts may be larger or smaller) but the impact of the calculation remains the same: it is an absurd conclusion. Galen's theory would require

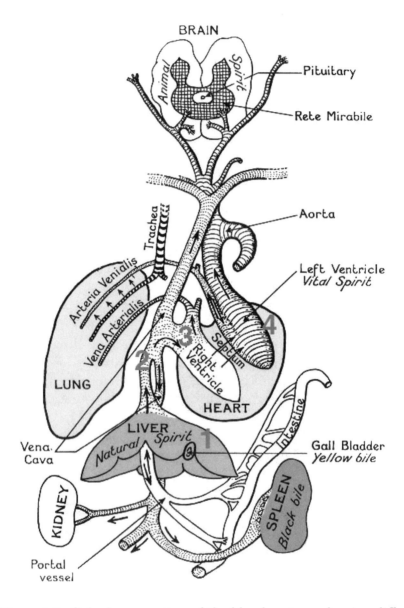

Figure 1.2: Galen's conception of the blood system, showing different "spirits" traveling in one direction but not circulating. Reproduced by permission, courtesy of Dr. Barbara Becker.

a human being to consume and produce a quantity of fluid many times the volume of the human body (about 100 liters) in a day! This is a physical impossibility, so the only possible conclusion in that Galen's model is wrong.

This led Harvey to propose the model that we know today: that blood is not consumed by the tissues but instead returns to the heart and is reused (Schultz 2002). This is why we call the heart and blood vessels part of the circulatory system of the body. This model was controversial at the time—some people proclaimed they would "rather be wrong with Galen, than right with Harvey"— but eventually became accepted as the standard model. What is remarkable is that Harvey's argument, despite being grounded in empirical data, was strictly mathematical. He adopted the assumptions of Galen, made the calculations, and got a result that was inconsistent with reality. This is an excellent example of how mathematical modeling can be useful by providing clear evidence against a wrong hypothesis.

1.2 Parameters and variables in models

Many biologists remain skeptical of mathematical modeling. The criticism can be summarized like this: a theoretical model either agrees with experiment, or it does not. In the former case, it is useless, because the data are already known; in the latter case, it is wrong! As I indicated above, the goal of mathematical modeling is not to reproduce experimental data; otherwise, indeed, it would be of interest only to theoreticians. The correct question to ask is, does a theoretical model help us understand the real thing? There are at least three ways in which a model can be useful:

1. A model can help a scientist make sense of complex data by testing whether a particular mechanism explains the observations. Thus, a model can help clarify our understanding by throwing away the nonessential features and focusing on the most important ones.

2. A mathematical model can make predictions for situations that have not been observed. It is easy to change parameters

in a mathematical model and calculate the effects. This can lead to new hypotheses that can be tested by experiments.

3. Model predictions can lead to better experimental design. Instead of trying a whole bunch of conditions in an experiment, the theoretical model can first suggest which ones will produce big effects, and thus can save a lot of work for the lab scientist.

To make a useful model of a complex living system, you have to simplify it. Even if you are only interested in a part of it (for instance, a cell or a single molecule), you have to make simplifying choices. A small protein has thousands of atoms; a cell consists of millions of molecules, which all interact with one another. Keeping track mathematically of every single component is daunting, if not impossible. To build a useful mathematical model, you must choose a few quantities that describe the system sufficiently to answer the questions of interest. For instance, if the positions of a couple of atoms in the protein you are studying determine its activity, those positions would make natural quantities to include in your model. You will find more specific examples of models later in this chapter.

Once you have decided on the essential quantities to be included in the model, these are divided into *variables* and *parameters*. As suggested by the name, a variable typically varies over time, and the model tracks the changes in its value; parameters usually stay constant or change more slowly. However, that is not always the case. The most important difference is that variables describe quantities within the system being modeled, while parameters usually refer to quantities which are controlled by something outside the system.

As you can see from this definition, the same quantity can be a variable or a parameter, depending on the scope of the model. Let's go back to our example of modeling a protein: usually the activity (and the structure) of a protein is influenced by external conditions, such as pH and temperature; these would be natural parameters for a model of the molecule. However, if we model an entire organism, the pH (e.g., of the blood plasma) and temperature are controlled by physiological processes in the organism, and thus these quantities would then be considered variables.

1.2. PARAMETERS AND VARIABLES IN MODELS

Perhaps the clearest way to differentiate between variables and parameters is to think about how you would present the quantities visually. We discuss plotting graphs of functions in Chapter 2, and plotting data sets in Chapter 3, but you have likely seen many such plots before. Consider which of the quantities you would to plot to describe the system you are modeling. If the quantity belongs on either axis, it is a variable, since it requires a range of values to illustrate how it changes. The rest of the quantities can be called parameters. Of course, depending on the question you ask, the same quantity may be plotted on an axis or not, which is why this classification is not absolute.

After specifying the essential variables for the model, we can describe a complex and evolving biological system in terms of its *state*. This is a general term, but it usually means the values of all the variables chosen for the model, which are often called *state variables*. For instance, an ion channel can be described with the state variable of conformation, which may be in a open state or in a closed state. The range, or collection of all different states of the system, is called the *state space* of the model. Below are examples of models of biological systems with diverse state spaces.

1.2.1 discrete state variables: genetics

Some genes are present in a population as two different versions, called *alleles*—let us use letters A and B to label them. One may describe the genetic state of an individual based on which allele it carries. If this individual is haploid (e.g., a bacterium), then it only carries a single copy of the genome, and its state can be described by a single variable with the state space of A or B.

A diploid organism (e.g., a human) possesses two copies of each gene (unless the gene is on one of the sex chromosomes, X or Y); each copy may be in either state A or B. This may seem to suggest that there are four different values in the genetic state space, but if the order of the copies does not matter (which is usually the case), then AB and BA are effectively the same, so the state space consists of three values: AA, BB, and AB.

1.2.2 continuous state variables: concentration

Suppose that a biological molecule is produced at a certain rate and degraded at a different rate, and we would like to describe the quantity of the molecule, usually expressed as a concentration. The relevant variables here are concentration and time (you will see those variables on the axes of many plots in biochemistry.) Concentration is the ratio of the number of molecules and the volume, so the state space can be any positive real number (although practically speaking, there is a limit on how many molecules can fit inside a given volume, but for simplicity we can ignore this).

Going even further, let us consider an entire cell, which contains a large number of different molecules. We can describe the state of a cell as the collection of all the molecular concentrations, with the parameters being the rates of all the reactions going on among those molecules. The state space for this model with N different molecules is N positive real numbers.

Discussion questions:

For the biological models described below, divide the quantities into variables and parameters, and specify the state space of the model. Note that there may be more than one correct interpretation, so explain your decision in terms of the questions that you would like to ask of the model.

Discussion 1.2.1. The volume of blood pumped by the heart during a certain amount of time, depending on the heart rate and the ejection volume.

Discussion 1.2.2. The number of wolves in a national forest, depending on the number of wolves in the previous year, the birth rate, the death rate, and the migration rate.

Discussion 1.2.3. The fraction of hemes in hemoglobin (a transport protein in red blood cells) that are bound to oxygen, depending on the partial pressure of oxygen and the binding cooperativity of hemoglobin.

Discussion 1.2.4. The number of mutations that occur in a genome, depending on the mutation rate, the amount of time, and the length of the genome.

Discussion 1.2.5. The concentration of a drug in the blood stream, depending on the dose, time after administration, and the rate of metabolism (processing) of the drug.

Discussion 1.2.6. Describe an outbreak of an infectious disease in a city (described in terms of the fractions of infected, healthy, and recovered people), depending on the rate of infection, rate of recovery, and the mortality rate of the disease.

1.3 First steps in R programming

A central goal of this book is to help you gain experience with computation, which requires learning some programming (cool kids call it "coding"). Programming is a way of interacting with computers through a symbolic language, unlike the graphic user interfaces that we're all familiar with. Basically, programming allows you to make a computer do exactly what you want it to do.

There is a vast number of computer languages with distinct functionalities and personalities. Some are made to talk directly to the computer's "brain" (CPU and memory), such as assembly, while others are better suited for human comprehension (e.g., python or Java). Programming in any language involves two parts: (1) writing a program (code) using the commands and the syntax for the language and (2) running the code by using a compiler or interpreter to translate the commands into machine language and then making the computer execute the actions. If your code has a mistake in it, the compiler or interpreter should catch it and return an *error message* to you instead of executing the code. Sometimes, though, the code may pass muster with the interpreter/compiler, but it may still have a mistake (bug). This can be manifested in two different ways: either the code execution does not produce the result that you intended, or it hangs up or crashes the computer (the latter is hard to do with the kind of programming we will be doing). We will discuss errors and how to prevent and catch these bugs as you develop your programming skills.

In this course, our goal is to compute mathematical models and to analyze data, so we choose a language that is designed for these tasks, which is called R. To proceed, you'll need to download and install R, which is freely available at cran.r-project.org/. In addition to downloading the language (which includes the interpreter that allows you to run R code on your computer), you need to download a graphic interface for writing, editing, and running R code, called R Studio (coders call this an IDE, or an Integrated Developer Environment), which is also free and is available at www.rstudio.com/products/rstudio/download/.

1.3.1 numbers and arithmetic operations

When you get down to the brass tacks, all computation rests on performing *arithmetic operations*: addition, subtraction, multiplication, division, exponentiation, and so forth. The first thing you will do is to learn how to perform arithmetic operations in R. Open R Studio, and you will see a window divided into four frames. The bottom left frame contains what is called the *prompt*, where you can type a command, hit return, and it will be executed by the R interpreter. For example, type in 5-3 on the R prompt and hit return, and this is what you will see:

```
> 5-3

[1] 2
```

You see that R returns the result by printing it out on the screen. The number in square brackets [1] is not important for now; it is useful when the answer contains many numbers and has to be printed out on many rows. The second number is the result of the calculation.

The symbols used for arithmetic operations are what you'd expect: +, -, *, / are the four standard operations, and ^ is the symbol for exponentiation. For example, type 2^3 on the R command prompt, and you should see:

1.3. FIRST STEPS IN R PROGRAMMING

```
> 2^3
```

```
[1] 8
```

For numbers that are either very large or very small, it's too cumbersome to write out all the digits, so R, like most computational platforms, uses the *scientific notation*. For instance, if you want to represent 1.4 billion, you type in the following command (note that 10 to the ninth power is represented as e+09, and the prefix 1.4 is written without any multiplication sign):

```
> 1.4*10^9
```

```
[1] 1.4e+09
```

There are also certain numbers built into the R language, most notably π and e, which can be accessed as follows:

```
> pi
```

```
[1] 3.141593
```

```
> exp(1)
```

```
[1] 2.718282
```

The expression `exp()` is an example of a *function*, which we discuss in section 2.3; it returns the value of e raised to the power of the number in parenthesis; hence, `exp(1)` returns e. Notice that although both numbers are irrational, and thus have infinitely many decimal digits, R only prints out a few of them. This doesn't mean that it doesn't have more digits in memory, but it only displays a limited number to avoid clutter. The number of digits to be displayed can be changed; for example, to display 10 digits, type in `options(digits=10)`.

Computers are very good at computation, as their name suggests, but they have limitations. To manipulate numbers, they must be stored in computer memory, but computer memory is finite. There is a limit to the length of the number that is feasible to store on a computer. This has implications for both very large numbers and very small numbers that are close to zero, because both require many digits for storage.

All programming languages have an upper limit on the biggest number that can be stored and worked with. If an arithmetic operation results in a number larger than that limit, the computer will call it an *overflow* error. Depending on the language, this may stop the execution of the program, or else produce a non-numerical value, such as "NaN" (not a number) or "Inf" (infinite). Do exercise 1.3.3 to investigate the limitations of R for large numbers.

Very small numbers present their own challenges. As with very large numbers, a computer cannot store an arbitrary number of digits after the decimal (or binary) point. Therefore, there is also the smallest number that a programming language will accept and use, and storing a smaller number produces an *underflow* error. This will either cause the program execution to stop or to return the value 0 instead of the correct answer. Do exercise 1.3.4 to investigate the limitations of R for small numbers.

This last fact demonstrates that computer operations are approximate, as they are limited by what's called the *machine precision*, which is illustrated in exercise 1.3.5. For instance, two similar numbers, if they are within the machine precision of each other, will be considered the same by the computer. Modern computers have large memories, and their machine precision is very good, but sometimes this error presents a problem (e.g., when subtracting two numbers). A detailed discussion of machine error is beyond the scope of this text, but anyone performing computations must be aware of its inherent limitations.

Programming exercises:

Exercise 1.3.1. Calculate the value of π raised to the tenth power.

Exercise 1.3.2. Use the scientific notation to multiply four billion by π.

Exercise 1.3.3. Use the scientific notation with large exponents (e.g., 10^{100}, 10^{500}) to find out what happens when you give R a

1.3. FIRST STEPS IN R PROGRAMMING

number that is too large for it to handle. At approximately what order of magnitude does R produce an overflow error?

Exercise 1.3.4. In the same fashion, find out what happens when you give R a number that is too small for it to handle. At approximately what order of magnitude does R produce an underflow error?

Exercise 1.3.5. How close can two numbers be before R treats them as identical? Subtract two numbers that are close to each other, like 24 and 24.001, and keep making them closer to each other, until R returns a difference of zero. Report at what value of the actual difference this happens.

1.3.2 variable assignment

Variables in programming languages are used to store and access numerical or other information. After *assigning* it a value for the first time (*initializing*), a variable name can be used to represent the value assigned to it. Invoking the name of variable recalls the stored value from the computer's memory. There are a few rules about naming variables: a name cannot be a number or an arithmetic operator like +; in fact, it cannot contain symbols for operators or spaces inside the name, or else confusion would reign. Variable names may contain numbers, but not as the first character. When writing code, it is good practice to give variables informative names, like `height` or `city.population`.

The symbol "=" is used to assign a value to a variable in most programming languages, and can be used in R too. However, it is customary for R to use the symbols "<-" together to indicate assignment, like this:

```
> var1 <- 5
```

After this command, the variable `var1` has the value 5, which you can see in the top right frame in R Studio called "Environment." To display the value of the variable as an output on the screen, use the special command `print()` (it's actually a *function*, which we discuss in Chapter 2). The following two commands show that the value of a variable can be changed after it has been initialized:

```
> var1 <- 5
> var1 <- 6
> print(var1)

[1] 6
```

While seemingly contradictory, the commands are perfectly clear to the computer: first `var1` is assigned the value 5, and then it is assigned 6. After the second command, the first value is overwritten, so any operations that use the variable `var1` will be using the value of 6.

Entire expressions can be placed on the right–hand side of an assignment command: they could be arithmetic or logical operations as well as functions. For example, the following commands result in the value 6 being assigned to the variable `var2`:

```
> var1 <- 5
> var2 <- var1+1
> print(var2)

[1] 6
```

Even more mind-blowing is that the same variable can be used on both sides of an assignment operator! The R interpreter first looks on the right-hand side to evaluate the expression and then assigns the result to the variable name on the left-hand side. So for instance, the following commands increase the value of `var1` by 1, and then assign the product of `var1` and `var2` to the variable `var2`:

```
> var1 <- var1 + 1
> print(var1)

[1] 6

> var2 <- var1-1
> print(var2)
```

1.3. FIRST STEPS IN R PROGRAMMING 25

```
[1] 5

> var2 <- var1*var2
> print(var2)

[1] 30
```

We have seen examples of how to assign values to variables, so here is an example of how NOT to assign values, with the resulting error message:

```
> var1 + 1 <- var1

Error in var1 + 1 <- var1:  could not find function "+<-"
```

The left-hand side of an assignment command should contain only the variable to which you are assigning a value, not an arithmetic expression to be performed.

Programming exercises:

The following commands or scripts do not work as intended. Find the errors and correct them, then run them to make sure they do what they are intended to do.

Exercise 1.3.6. Add 5 and 3 and save it into variable my.number
```
5 + 3 <- my.number
```

Exercise 1.3.7. print the value of my.number on the screen
```
print[my.number]
```

Exercise 1.3.8. replace the value of my.number with 5 times its current value
```
my.number <- 5my.number
```

Exercise 1.3.9. assign the values of 7 and 8 to variables a and b, respectively; multiply them; and save the results in variable x
```
a<-7
b<-8
x<-ab
print(x)
```

1.4 Computational projects

In this part you will translate mathematical models into R commands and make the computer perform calculations. The first step is to give descriptive names to these variables (instead of single letters!), and then use these variables to calculate the predictions of the models. Units cannot be attached to variable values but it is a good idea to specify them in your code by using to add a comment (e.g., `m <- 100 # kg`).

1. Heart pumping with a constant rate, where V_{tot} is the total volume of blood pumped by the heart over time, V_s is the stroke volume, R is the heart rate and t is time:

$$V_{tot} = V_s R t$$

2. Bacterial population that doubles every hour, where P is the current population, P_0 is the initial population, and t is the elapsed time:

$$P = P_0 2^t$$

3. The rate of an enzyme-catalyzed reaction, where v is the rate of the reaction, v_{max} is the maximum reaction rate, K_M is the Michaelis constant, and A is the concentration of the substrate:

$$v = \frac{v_{max} A}{K_M + A}$$

Computational tasks

1. Implement the heart-pumping model: (a) create variables for the stroke volume, heart rate, and time (come up with your own descriptive names) by assigning them reasonable values (e.g. 80, 70, and 120, respectively); (b) define the variable for total volume as a calculation based on the variables you have defined. Calculate the volume of blood pumped by a heart beating at 80 beats per minute with stroke volume of 70 mL over 80 minutes (the length of one class).

1.4. COMPUTATIONAL PROJECTS

2. Implement the model of bacterial population: (a) create variables for the time, and initial population (come up with your own descriptive names) by assigning them reasonable values (e.g., 20 hours and 10^6 cells, respectively); (b) define the variable for bacterial population as a calculation based on the variables you had defined. Using this script, calculate the bacterial population after 3 days if the initial population is a million.

3. Implement the model for Michealis-Menten kinetics: 1) create variables for the substrate concentration, maximum reaction rate, the Michaelis constant (come up with your own descriptive names) by assigning them reasonable values (e.g., 1 mM, 20 1/s, 5 mM, respectively); 2) define the variable for reaction rate as a calculation based on the variables you had defined. Using this script, calculate the rate of the reaction given the substrate concentration of 1 mM, the maximum rate of 10 1/s, and the Michaelis constant of 30 mM.

Chapter 2

Functions and their graphs

> *Some fathers, if you ask them for the time of day, spit silver dollars.*
>
> —Donald Barthelme, *The Dead Father*

Mathematical models describe how various quantities affect one another. In Chapter 1, we learned that these descriptions can be written down, often in the form of an equation. For instance, we can describe the total volume of blood pumped over a period of time as the product of stroke volume, the heart rate and the number of minutes, which can be written as an equation. The different quantities have their own meaning and roles, depending on what they stand for. To better describe how these quantities are related, we use the deep idea of mathematical functions. In this chapter you will learn to do the following:

1. use dimensional analysis to deduce the meaning of quantities in a model;

2. understand the concept of function, and dependent and independent variables;

3. recognize basic functional forms and the shape of their graphs;

30   CHAPTER 2. FUNCTIONS AND GRAPHS

4. use R to plot functions; and

5. understand basic models of reaction rates.

2.1 Dimensions of quantities

What distinguishes a mathematical model from a mathematical equation is that the quantities involved have a real-world meaning. Each quantity represents a measurement, and associated with each one are the *units* of measurement. The number 173 is not enough to describe the height of a person—you are left to wonder, 173 what? meters, centimeters, nanometers, light-years? Obviously, only centimeters make sense as a unit of measurement for human height; but if we were measuring the distance between two animals in a habitat, meters would be a reasonable unit; and if it were the distance between molecules in a cell, we would use nanometers. Thus, any quantity in a mathematical model must have associated units, and any graphs of these quantities must be labeled accordingly.

In addition to units, each variable and parameter has a meaning, which is called the *dimension* of the quantity. For example, any measurement of length or distance has the same dimension, although the units may vary. The value of a quantity depends on the units of measurement, but its essential dimensionality does not. One can convert a measurement in meters to that in light-years or cubits, but one cannot convert a measurement in number of sheep to seconds—that conversion has no meaning.

This leads us to the fundamental rule of mathematical modeling: terms that are added or subtracted must have the same dimension. This gives mathematical modelers a useful tool called *dimensional analysis*, which involves replacing the quantities in an equation with their dimensions. This practice serves as a check that all dimensions match, as well as allowing us to deduce the dimensions of any parameters for which the dimension was not specified. (Smith 1968)

Example. As we saw in Chapter 1, the relationship between the amount blood pumped by a heart and time elapsed is expressed in the following equation, where V_{tot} and V_s are the total volume and stroke volume, respectively, R is the heart rate, and t is the time:

2.1. DIMENSIONS OF QUANTITIES

$$V_{tot} = V_s R t$$

The dimension of a quantity X is denoted by $[X]$; for example, if t has the dimension of time, we write $[t] = time$. The dimension of volume is $[V_{tot}] = length^3$, the dimension of stroke volume is $[V_s] = volume/beat$, and the dimension of time t is time, so we can rewrite the equation above in dimensional form:

$$length^3 = length^3/beat * R * time$$

Solving this equation for R, we find that it must have the dimensions of $[R] = beats/time$. It can be measured in beats per minute (typical for heart rate), or beats per second, beats per hour, and so forth but the dimensionality of the quantity cannot be changed without making the model meaningless.

There are also *dimensionless* quantities, or pure numbers, which are not tied to a physical meaning at all. Fundamental mathematical constants, like π or e, are classic examples, as are some important physical constants, like the Reynolds number in fluid mechanics (Strogatz 2001). Quantities with a dimension can be made dimensionless by dividing them by another quantity with the same dimension and "canceling" the dimensions. For instance, we can express the height of a person as a fraction of the mean height of the population; then the height of a tall person will become a number greater than 1, and the height of a short one will become less than 1. This new dimensionless height does not have units of length—the unit has been divided out by the mean height. This is known as *rescaling* the quantity by dividing it by a preferred scale. There is a fundamental difference between rescaling and changing the units of a quantity: when changing the units (e.g., from inches to centimeters), the dimension remains the same, but if one divides the quantity by a scale, it loses its dimension.

Example. The model for a population of bacteria that doubles every hour is described by the equation, where P_0 is initial number of bacteria and P is the population after t hours:

$$P = P_0 2^t$$

Let us define the quantity $R = P/P_0$, so we can say that population increased by a factor of R after t hours. This ratio is a dimensionless quantity because P and P_0 have the same dimension of bacterial population, which cancel out. The equation for R can be written as follows:

$$R = 2^t$$

According to dimensional analysis, both sides of the equation have to be dimensionless, so t must also be a dimensionless variable. This is surprising, because t indicates the number of hours the bacterial colony has been growing. This reveals the subtle fact that t is a rescaled variable obtained by dividing the elapsed time by the length of the reproductive cycle. Because of the assumption that the bacteria divide exactly once an hour, t counts the number of hours, but if they divided once a day, t would denote the number of days. So t doesn't have units or dimensions, but instead denotes the dimensionless number of cell divisions.

Math exercises:

For each biological model below, determine the dimensions of the parameters, based on the given dimensions of the variables.

Exercise 2.1.1. Model of number of mutations M as a function of time t:
$$M(t) = M_0 + \mu t$$

Exercise 2.1.2. Model of molecular concentration C as a function of time t:
$$C(t) = C_0 e^{-kt}$$

Exercise 2.1.3. Model of tree height H (length) as a function of age a (time):
$$H(a) = \frac{ba}{c+a}$$

Exercise 2.1.4. Model of cooperative binding of ligands, with fraction of bound receptors θ as a function of ligand concentration L:
$$\theta(L) = \frac{L^n}{L^n + K_d}$$

Exercise 2.1.5. Model of concentration of a gene product G (concentration) as a function of time t:

$$G(t) = G_m(1 - e^{-\alpha t})$$

Exercise 2.1.6. Michaelis-Menten model of enzyme kinetics, where v is reaction rate (1/time) and S is substrate concentration:

$$v(S) = \frac{v_{max}S}{K_m + S}$$

Exercise 2.1.7. Logistic model of population growth, where P is population size and time t:

$$P(t) = \frac{Ae^{kt}}{1 + B(e^{kt} - 1)}$$

2.2 Functions and their graphs

A relationship between two variables addresses the basic question: when one variable changes, how does this affect the other? An equation, like the examples given in section 2.1, allows one to calculate the value of one variable based on the other variable and parameter values. In this section we seek to describe more broadly how two variables are related by using the mathematical concept of functions.

Definition 2.1. A *function* is a mathematical rule which has an *input* and an *output*. A function returns a well-defined output for every input; that is, for a given input value the function returns a unique output value.

This is an abstract definition of a mathematical function; it doesn't have to be expressed as an algebraic equation, it only has to return a unique output for any given input value. In mathematics courses, we usually write them down in terms of algebraic expressions. As in mathematical models, you will see two different kinds of quantities in equations that define functions: variables and parameters. The input and the output of a function may both be variables, in which case the input is called the *independent variable* and the output is called the *dependent variable*.

The relationship between the input and the output can be illustrated in a *graph*, which is a collection of paired values of the independent and dependent variable drawn as a curve in the plane. Although it shows how the two variables change relative to each other, parameters may change too, which results in a different graph of the function. While graphing calculators and computers can draw graphs for you, it is very helpful to have an intuitive understanding about how a function behaves and how the behavior depends on the parameters. Here are three questions that can help you picture the relationship (assume x is the independent variable and it is a nonnegative real number):

1. What is the value of the function at $x = 0$?

2. What does the function do when x becomes large ($x \to \infty$)?

3. What does the function do between the two extremes?

Below are examples of fundamental functions used in biological models with descriptions of how their parameters influence their graphs.

2.2.1 linear and exponential functions

You are probably familiar with linear and exponential functions from algebra courses. However, they are so commonly used that it is worth going over them to refresh your memory and perhaps to see them from another perspective.

Definition 2.2. A *linear function* $f(x)$ is one for which the difference in two function values is the same for a specific difference in the independent variable.

In mathematical terms, this can be written an equation for any two values of the independent variable x_1 and x_2 and a difference Δx:

$$f(x_1 + \Delta x) - f(x_1) = f(x_2 + \Delta x) - f(x_2)$$

2.2. FUNCTIONS AND THEIR GRAPHS

The general form of the linear function is written as follows:

$$f(x) = ax + b \tag{2.1}$$

The function contains two parameters: the *slope a* and the *y-intercept b*. The graph of the linear function is a line (hence the name), and the slope a determines its steepness. A positive slope corresponds to a graph that increases as x increases, and a negative slope corresponds to a declining function. At $x = 0$, the function equals b, and as $x \to \infty$, the function approaches positive infinity if $a > 0$ and approaches negative infinity if $a < 0$.

Definition 2.3. An *exponential function* $f(x)$ is one for which the ratio of two function values is the same for a specific difference in the independent variable.

Mathematically speaking, this can be written as follows for any two values of the independent variable x_1 and x_2 and a difference Δx:

$$\frac{f(x_1 + \Delta x)}{f(x_1)} = \frac{f(x_2 + \Delta x)}{f(x_2)}$$

Exponential functions can be written using different symbolic forms, but they all have a constant base with the variable x in the exponent. I prefer to use the constant e (base of the natural logarithm) as the base of all exponential functions, for reasons that will become apparent in Chapter 15. This does not restrict the range of possible functions, because any exponential function can be expressed using base e, using a transformation: $a^x = e^{x \ln(a)}$. So let us agree to write exponential functions in the following form:

$$f(x) = ae^{rx} \tag{2.2}$$

The function contains two parameters: the *rate constant r* and the *multiplicative constant a*. The graph of the exponential function is a curve that crosses the y-axis at $y = a$ (plug in $x = 0$ to see that this is the case). As x increases, the behavior of the graph depends on the sign of the rate constant r. If $r > 0$, the function approaches infinity (positive if $a > 0$, negative if $a < 0$) as $x \to \infty$. If $r < 0$, the function decays at an ever-decreasing pace and asymptotically approaches zero as $x \to \infty$. Thus the graph of $f(x)$ is a curve either going to infinity or a curve asymptotically approaching 0, and the steepness of the growth or decay is determined by r.

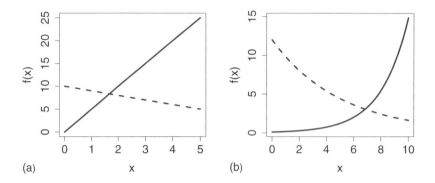

Figure 2.1: Plots of two linear functions (a) and two exponential functions (b). Can you identify which linear function has the positive slope and which one negative? Which exponential function has a positive rate constant and which a negative one?

Math exercises:

Answer the questions below, some of which refer to the function graphs in Figure 2.1.

Exercise 2.2.1. Which of the linear graphs in the first figure corresponds to $f(x) = 5x$ and which corresponds to $f(x) = 10 - x$? State which parameter allows you to connect the function with its graph and explain why.

Exercise 2.2.2. Which of the exponential graphs in the second figure corresponds to $f(x) = 0.1e^{0.5x}$ and which corresponds to $f(x) = 12e^{-0.2x}$? State which parameter allows you to connect the function with its graph and explain why.

Exercise 2.2.3. Demonstrate algebraically that a linear function of the form given in equation 2.1 satisfies the property of linear functions given in definition 2.2.

2.2. FUNCTIONS AND THEIR GRAPHS

Exercise 2.2.4. Demonstrate algebraically that an exponential function of the form given in equation 2.2 satisfies the property of exponential functions given in definition 2.3.

Exercise 2.2.5. Modify the exponential function by adding a constant term to it: $f(x) = ae^{rx} + b$. What is is the value of this function at $x = 0$?

Exercise 2.2.6. How does the function defined in the previous exercise, $f(x) = ae^{rx} + b$, behave as $x \to \infty$ if $r > 0$?

Exercise 2.2.7. How does the function $f(x) = ae^{rx} + b$ behave as $x \to \infty$ if $r < 0$?

2.2.2 rational and logistic functions

Let us now turn to more complex functions, made up of simpler components that we understand. Consider a ratio of two polynomials, called a *rational function*. The general form of such functions can be written down as follows, where ellipsis stands for terms with powers lower than n or m:

$$f(x) = \frac{a_0 + \cdots + a_n x^n}{b_0 + \cdots + b_m x^m} \qquad (2.3)$$

The two polynomials may have different degrees (highest power of the terms, n and m), but they are usually the same in most biological examples. The reason is that if the numerator and the denominator are "unbalanced," one will inevitably overpower the other for large values of x, which would lead to the function either increasing without bound to infinity (if $n > m$) or decaying to zero (if $m > n$). There's nothing wrong with that, mathematically, but rational functions are most frequently used to model quantities that approach a nonzero asymptote for large values of the independent variable.

For this reason, let us assume $m = n$ and consider what happens as $x \to \infty$. All terms other than the highest-order terms become very small in comparison to x^n (this is something you can demonstrate for yourself using R), and thus both the numerator and the denominator approach the terms with power n. This can be written using the mathematical limit notation $\lim_{x \to \infty}$, which describes the value that a function approaches when the independent variable increases without bound:

$$\lim_{x \to \infty} \frac{a_0 + \cdots + a_n x^n}{b_0 + \cdots + b_n x^n} = \frac{a_n x^n}{b_n x^n} = \frac{a_n}{b_n}$$

Therefore, the function approaches the value of a_n/b_n as x grows.

Similarly, let us consider what happens when $x = 0$. Plugging this into the function results in all of the terms vanishing except for the constant terms, so

$$f(0) = \frac{a_0}{b_0}$$

Between 0 and infinity, the function either increases or decreases monotonically, depending on which value (a_n/b_n or a_0/b_0) is greater. Two examples of plots of rational functions are shown in the right panel of Figure 2.2, which shows graphs increasing from 0 to 1. Depending on the degree of the polynomials in a rational function, it may increase more gradually (solid line) or in a more step-like fashion (dashed line).

Example. The following model, called the *Hill equation*, describes the fraction of receptor molecules which are bound to a ligand, which is a chemical term for a free molecule that binds to another, typically larger, receptor molecule. θ is the fraction of receptors bound to a ligand, L denotes the ligand concentration, K_d is the dissociation constant, and n called the *binding cooperativity* or Hill coefficient:

$$\theta = \frac{L^n}{L^n + K_d}$$

The Hill equation is a rational function, and Figure 2.2 shows plots of the graphs of two such function in the right panel. This model is further explored in exercise 2.2.10.

A similar yet distinct category of functions, commonly seen in population models are called *logistic*. There are variations on how they are written down, but here is one general form:

$$f(x) = \frac{ae^{rx}}{b + e^{rx}} \tag{2.4}$$

2.2. FUNCTIONS AND THEIR GRAPHS

The numerator and denominator both contain exponential functions with the same power. If $r > 0$, then when $x \to \infty$, the denominator approaches e^{rx}, since it becomes much greater than b, and we can calculate:

$$\lim_{x \to \infty} \frac{ae^{rx}}{e^{rx}} = a; \text{ if } r > 0$$

In contrast, if $r < 0$, then the numerator approaches zero as $x \to \infty$, and so does the function:

$$\lim_{x \to \infty} = \frac{0}{b} = 0; \text{ if } r < 0 \ .$$

Notice that switching the sign of r has the same effect as switching the sign of x, since they are multiplied. Which means that for positive r, if x is extended to negative infinity, the function approaches 0. This is illustrated in the right panel in Figure 2.2, which shows two logistic functions increasing from 0 to a positive level, one with $a = 20$ (solid line) and the second with $a = 10$ (dashed line). The graph of logistic functions has a characteristic *sigmoidal* (S-shaped) shape, and its steepness is determined by the rate r: if r is small, the curve is soft; if r is large, the graph resembles a step function.

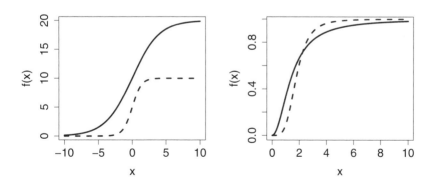

Figure 2.2: Examples of two graphs of logistic functions (left) and two Hill functions (right).

Math exercises:

For each biological model described here, answer the following questions in terms of the parameters in the models, assuming all are non-negative real numbers. (1) What is the value of the function when the independent variable is 0? (2) What value does the function approach when the independent variable goes to infinity? (3) Verbally describe the behavior of the functions between 0 and infinity (e.g., function increases, decreases).

Exercise 2.2.8. Model of number of mutations M as a function of time t:
$$M(t) = M_0 + \mu t$$

Exercise 2.2.9. Model of molecular concentration C as a function of time t:
$$C(t) = C_0 e^{-kt}$$

Exercise 2.2.10. Model of cooperative binding of ligands, with fraction of bound receptors θ as a function of ligand concentration L:
$$\theta = \frac{L^n}{L^n + K_d}$$

Exercise 2.2.11. Model of tree height H (length) as a function of age a (time):
$$H(a) = \frac{ba}{c + a}$$

Exercise 2.2.12. Model of concentration of a gene product G (concentration) as a function of time t:
$$G(t) = G_m(1 - e^{-at})$$

Exercise 2.2.13. Michaelis-Menten model of enzyme kinetics, where v is reaction rate (1/time) and S is substrate concentration:
$$v(S) = \frac{v_{max} S}{K_m + S}$$

Exercise 2.2.14. Logistic model of population growth, where P is population size and t is time:
$$P(t) = \frac{A e^{kt}}{1 + B(e^{kt} - 1)}$$

2.3 Scripts, functions, and plotting in R

2.3.1 writing scripts and calling functions

Programming means arranging a number of commands in a particular order to perform a task. It would be ludicrous to type them one at a time into the command line. Instead, the commands are written into a file called a program or script (the name depends on the type of language; since R is a scripting language, you will be writing scripts), which can be edited, saved, copied, and so forth. To open a new script file, in R Studio, go to File menu, and choose New R Script. This will open an editor window, where you can type your commands. To save the script file (do this often!), click the Save button (with the little floppy disk icon) or select Save from the File menu. You will also see small buttons at the top of the window that read "Run," "Re-run," and "Source." The first two will run either the current line or a selected region of the script, while the Source button will run the entire file. Now that you know how to create a script, you should never type your R code into the command line, unless you're testing a single command to see what it does or are looking up help.

R comes equipped with many functions that correspond to standard mathematical functions. As seen section 1.3, `exp()` is the exponential function that returns e raised to the power of the input value. Other common functions are: `sqrt()` returns the square root of the input value; and `sin()` and `cos()` return the sine and the cosine of the input value, respectively. Note that all these function names are followed by parentheses, which is a hallmark of a function (in R as well as in mathematics). This indicates that the input value has to go there, for example, `exp(5)`. To compute the value of e^5, save it into a variable called `var1`, and then print out the value on the screen, you can create the following script:

```
var1 <- exp(5)
print(var1)

## [1] 148.4132
```

The output of scripts here and elsewhere in the text is indicated by two hashtags to distinguish it from the script commands. Copy and paste the commands into a script file, save it, and then run it. You will see two things happen in R Studio: a variable named x appears in the Environment window (top right) with the value 148.41..., and the same value is printed out in the command line window (bottom left).

The most important principle of the procedural brand of programming (which includes R) is this: the computer (that is, the compiler or interpreter) evaluates the commands from top to bottom, one at a time. The variables are used with the values that they are currently assigned. If one variable (var1) was defined in terms of another (var2), and then var2 is changed later, this does not change the value of var2. Here is an illustration of how this works:

```
var2 <- 20
var1 <- var2/20
print(var2)
```

```
## [1] 20
```

```
var2 <- 10
print(var1)
```

```
## [1] 1
```

Notice that var1 doesn't change, because the R interpreter reads the commands one by one and does not go back to reevaluate the assignment for var1 after var2 is changed. Learning to think in this methodical, literal manner is crucial for developing programming skills.

2.3.2 vector variables

Variables are not restricted to single numbers. You can also store multiple numbers in a single variable, which is then called a vector. There are several ways of producing a vector of numbers in R. The first is to put together several numbers by listing them inside the function c():

2.3. SCRIPTS, FUNCTIONS, AND PLOTTING IN R

```
my.vec <- c(pi, 45, 912.8, 0)
print(my.vec)

## [1]    3.141593   45.000000 912.800000    0.000000
```

The variable `my.vec` is now a vector variable that contains four different numbers. Each of those numbers can be accessed individually by referencing its position in the vector, called the *index*. In the R language the the index for the first number in a vector is 1, the index for the second number is 2, and so forth. The index is placed in square brackets after the vector name, as follows:

```
print(my.vec[1])

## [1] 3.141593

print(my.vec[2])

## [1] 45

print(my.vec[3])

## [1] 912.8

print(my.vec[4])

## [1] 0
```

Another way to generate a sequence of numbers in a particular order is to use the colon operator, which produces a vector of integers from the first number to the last, inclusive. Here are two examples:

```
my.vec1 <- 1:20
print(my.vec1)

##  [1]  1  2  3  4  5  6  7  8  9 10 11 12 13 14 15
## [16] 16 17 18 19 20
```

```
my.vec2 <- 0:-20
print(my.vec2)

##  [1]   0  -1  -2  -3  -4  -5  -6  -7  -8  -9 -10
## [12] -11 -12 -13 -14 -15 -16 -17 -18 -19 -20
```

You can also access some, but not all, values stored in a vector simultaneously. To do this, enter a vector of positive integers inside the square brackets, either using the colon operator or using the c() function. Here are two examples: the first prints out the fourth through the tenth element of the vector my.vec1; the second prints out the first, fifth and eleventh elements of the vector my.vec2:

```
print(my.vec1[4:10])

## [1]  4  5  6  7  8  9 10

print(my.vec2[c(1, 5, 11)])

## [1]   0  -4 -10
```

If you want to generate a sequence of numbers with a constant difference other than 1, you're in luck: R provides a function called seq(). It takes three inputs: the starting value, the ending value, and the step (difference between successive elements). For example, to generate a list of numbers starting at 20 up to 50, with a step size of 3, type the first command; to obtain the same sequence in reverse, use the second command:

```
my.vec1 <- seq(20, 50, 3)
print(my.vec1)

##  [1] 20 23 26 29 32 35 38 41 44 47 50

my.vec2 <- seq(50, 20, -3)
print(my.vec2)

##  [1] 50 47 44 41 38 35 32 29 26 23 20
```

2.3. SCRIPTS, FUNCTIONS, AND PLOTTING IN R

2.3.3 arithmetic with vector variables

One of the advantages of vector variables is that you can perform operations on all the numbers stored in the vector using only one command. For instance, to multiply every element of the vector by the same number, it's enough to do the following:

```
NewVec <- 2 * my.vec
print(NewVec)

## [1]     6.283185    90.000000  1825.600000
## [4]     0.000000
```

You can also perform calculations with multiple vector variables, but this requires extra care. R can perform any arithmetic operation with two vector variables; for instance, adding two vectors results in a vector containing the sum of corresponding elements of the two vectors:

```
my.vec1 <- 1:5
my.vec2 <- 0:4
print(my.vec1)

## [1] 1 2 3 4 5

print(my.vec2)

## [1] 0 1 2 3 4

sum.vec <- my.vec1 + my.vec2
print(sum.vec)

## [1] 1 3 5 7 9
```

Take care that the two vectors have the same number of elements (length). If you try to operate on (e.g., add) two vectors of different lengths, R will return a warning and the result will not be what you expect:

```
my.vec1 <- 1:2
my.vec2 <- 0:4
print(my.vec1)
```

```
## [1] 1 2
```

```
print(my.vec2)
```

```
## [1] 0 1 2 3 4
```

```
sum.vec <- my.vec1 + my.vec2
```

```
## Warning in my.vec1 + my.vec2: longer object length is not
a multiple of shorter object length
```

```
print(sum.vec)
```

```
## [1] 1 3 3 5 5
```

The warning message tells you that the lengths of vectors don't match (more precisely, that one is not a multiple of the other), but R interprets your request, by adding my.vec1 (1, 2) to the first two elements of my.vec2, then adding my.vec1 to the third and forth element of my.vec2, and then finally adding the first element of my.vec1 to the last element of my.vec2, which is why sum.vec2 contains the numbers you see. To avoid these complications, when adding, subtracting, multiplying, or doing anything else to multiple vectors, make sure they have the same length.

Programming exercises:

Find errors in the following R scripts and correct them, then run them in R to make sure they do what they are intended to do.

Exercise 2.3.1. create a vector vec1 of ten integers and print the second and the eighth elements
```
vec1 <- 11:20
print(vec1[2:8])
```

2.3. SCRIPTS, FUNCTIONS, AND PLOTTING IN R   47

Exercise 2.3.2. create a vector vec2 of 7 integers, and print out 2 raised to the power of each of the elements
```
vec1 <- 0:6
print(2*vec2)
```

Exercise 2.3.3. create a vector vec1 and a vector vec2, add them and assign the result to vector sum.vec, and print out the third element of sum.vec
```
vec1 <- 0:5
vec2 <- 3:8
sum.vec <- vec1+vec2
print(sum.vec(3))
```

Exercise 2.3.4. create a vector vec1 and then multiply all of its elements by 20 and assign it to another vector
```
vec1<-seq(-3,2,0.1)
vec2 <- 2vec1
```

Exercise 2.3.5. create a vector vec1 and a vector vec2, and print out all the elements of the first divided by the second
```
vec1 <- 0:5
vec2 <- 3:8
print[vec1/vec2]
```

2.3.4 plotting graphs

There are several ways of creating plots of mathematical functions or data in R. If you want to plot a mathematical function, the simplest function is curve(). You can tell that this is a function, because it uses parentheses; the first input is an expression for the function, and the next two define the range of the independent variable over which to plot the graph. Two examples of plotting a quadratic function over the range 0–5 and an exponential variation over the range 0–10 are shown in Figure 2.3.

You can change the default look of the plot produced by curve by setting different options, which are optional inputs to curve. One option is the line width lwd, which can be increased from the default value of 1 to produce thicker curves, as demonstrated in the

```
curve(x^2, 0, 10, lwd = 3, xlab = "x",
    ylab = "quadratic",
    cex.axis = 1.5, cex.lab = 1.5)
curve(20 * exp(-0.5 * x), 0, 5, lwd = 3, xlab = "x",
    ylab = "exponential", cex.axis = 1.5, cex.lab = 1.5)
```

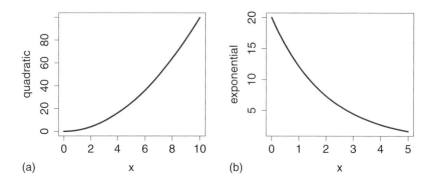

Figure 2.3: Two examples of plots using curve: (a) quadratic ($y = x^2$) and (b) exponential ($y = 20e^{-0.5x}$). R script for plots is shown above.

example. You can add labels on the x-and y-axes with xlab and ylab options, respectively; note that these are strings of characters, and thus must be put in quotes to differentiate them from a variable name. There is one very important option not shown in the example: that of overlaying a curve on top of an existing plot, which is done by typing add=TRUE. This option takes logical (Boolean) values TRUE and FALSE, which must be typed in all caps and without quotes.

In addition to curve, you can use the plot() function in R to create two-dimensional graphs from two vector-valued variables of the same length, for example, plot(x,y). This is also a function; its first input variable corresponds to the independent variable (e.g., x), which is plotted on the x-axis, and the second variable corresponds to the dependent variable (e.g., y), which is plotted on the y-axis.

2.3. SCRIPTS, FUNCTIONS, AND PLOTTING IN R 49

Figure 2.4 shows graphs of exponential and logistic function plotted using `plot`. The default plot style in R uses circles to indicate each plotted point. To change it, you need to set the option `t` (type), for example, setting `t='l'` (the lowercase letter L) produces a continuous line connecting the individual data points. The plot function has many options that you can change to determine the color, style, and other attributes of the plot.

You may also want to plot multiple graphs on the same figure. In R, each plot command produces a new plot window, so consider a new function called `lines()`, which overlays lines-style plots on top of an existing one. Let us illustrate this by plotting two different exponential functions on one plot and two different logistic functions on the second one, which were discussed in section 2.2. When you've got multiple plots on the same figure, they need to be distinct and labeled. To distinguish them, below I use the option `col` to specify the color of the plot, and I add a legend describing the parameters of each plot to Figure 2.4. The function has a lot of options, so if you want to understand the details, type `help(legend)` in the prompt or go to the Help tab in the lower right frame of R Studio and type "legend."

Programming exercises:

Find errors in the following R scripts and correct them. Run them in R to make sure they do what they are intended to do.

Exercise 2.3.6. plot a quadratic function with specified coefficients a, b, c over a given range of independent variable x
```
a<-10
b<- -15
c<-5
y<-a*x^2+b*x+c
x<-seq(-0.5,2,0.01)
plot(x,y,type='l')
```

Exercise 2.3.7. overlay two different plots of the logistic function with different values of the parameter r

```
x <- seq(0, 10, 0.5)
y <- 10 + 20 * exp(-0.5 * x)
plot(x, y, xlab = "x", ylab = "exponential", col = 1,
    lwd = 3)
y <- 10 + 20 * exp(-2 * x)
lines(x, y, col = 2, lwd = 3)
leg.txt <- c("b=10,a=20,r=-0.5", "b=10,a=20,r=-2")
legend("topright", leg.txt, col = 1:2, pch = c(1, NA),
    lty = c(0, 1), lwd = 3)
x <- seq(-10, 10, 1)
y <- 20 * exp(0.5 * x)/(1 + exp(0.5 * x))
plot(x, y, , xlab = "x", ylab = "logistic", col = 4,
    lwd = 3)
y <- 20 * exp(1.5 * x)/(1 + exp(1.5 * x))
lines(x, y, col = 2, lwd = 3)
leg.txt <- c("a=20,b=1,r=0.5", "a=20,b=1,r=1.5")
legend("topleft", leg.txt, col = c(4, 2), pch = c(1,
    NA), lty = c(0, 1), lwd = 3)
```

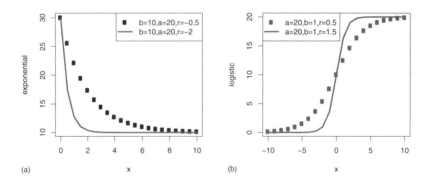

Figure 2.4: Overlaying multiple plots in R: (a) two exponential functions of the form $y = b + ae^{rx}$; (b) two logistic functions of the form $y = ae^{rx}/(b + e^{rx})$.

```
time<-0:100
a<-1000
b<-50
r<-0.1
Population<-a*exp(r*time)/(b+exp(r*time))
plot(time,Population,type='l')
r<-10
lines(time,Population,col=2)
```

2.4 Rates of biochemical reactions

Living things are dynamic: they change with time, and much of mathematical modeling in biology is interested in describing these changes. Some quantities change fast and others slowly, and every dynamic quantity has a rate of change, or *rate* for short. Usually, the quantity that we want to track over time is the variable, and to describe how it changes, we introduce a rate parameter. If we are describing changes over time, all rate parameters have dimensions with time in the denominator. As a simple example, the velocity of a physical object describes the change in distance over time, so its dimension is $[v] = length/time$.

Molecular reactions are essential for biology, whether they happen inside a bacterial cell or in the bloodstream of a human. *Reaction kinetics* refers to the description of the rates, or the speed, of chemical reactions. Different reactions occur at different rates, which may be dependent on the concentration of the reactant molecule. Consider a simple reaction of molecule A turning into molecule B, which is usually written by chemists with an arrow:

$$A \xrightarrow{k} B$$

But how fast does the reaction happen? To write down a mathematical model, we need to define the quantities involved. First, we have the concentration of the molecule A, with dimensions of concentration. Second, we have the rate of reaction, let us call it v, which has dimension of concentration per time (just like velocity is length per time). How are the two quantities related?

Constant (zeroth-order) kinetics. In some circumstances, the reaction rate v does not depend on the concentration of the reactant molecule A. In that case, the relationship between the *rate constant* k and the actual rate v is:

$$v = k \tag{2.5}$$

Dimensional analysis insists that the dimension of k must be the dimension of v, or *concentration/time*. This is known as constant, or zeroth-order kinetics, and it is observed at concentrations of A when the reaction is at its maximum velocity: for example, ethanol metabolism by ethanol dehydrogenase in the human liver cannot proceed any faster than about 1 drink per hour.

First-order kinetics. In other conditions, it is easy to imagine that increasing the concentration of the reactant A will speed up the rate of the reaction. A simple relationship of this type is linear:

$$v = kA \tag{2.6}$$

In this case, the dimension of the rate constant k is $1/time$. This is called first-order kinetics, and it usually describes reactions when the concentration of A is small and there are plenty of free enzymes to catalyze more reactions.

Michaelis-Menten model of enzyme kinetics. However, if the concentration of the molecule is neither small nor large, we need to consider a more sophisticated model. An enzyme is a protein that catalyzes a biochemical reaction, and it works in two steps: first it binds the substrate (at which point it can still dissociate and float away), and then it actually catalyzes the reaction, which is usually practically irreversible (at least by this enzyme) and releases the product. The enzyme itself is not affected or spent, so it is free to catalyze more reactions. Denote the substrate (reactant) molecule by A, the product molecule by B, the enzyme by E, and the complex of substrate and enzyme AE. The classic chemical scheme that describes these reactions is

$$A + E \underset{k_{-1}}{\overset{k_1}{\rightleftharpoons}} AE \overset{k_2}{\rightarrow} E + B$$

2.5. COMPUTATIONAL PROJECTS

You could write three different kinetic equations for the three different arrows in this scheme. Michaelis and Menten (1913) used the simplifying assumptions that the binding and dissociation happen much faster than the catalytic reaction. Based on this assumption, they were able to write down an approximate but extremely useful Michaelis-Menten model of an enzymatic reaction:

$$v = \frac{v_{max} A}{K_M + A} \qquad (2.7)$$

Here v refers to the rate of the entire catalytic process, that is, the rate of production of B, rather than any intermediate step. The reaction rate depends both on the concentration of the substrate A and on the two constants v_{max}, called the maximum reaction rate, and the constant K_M, called the Michaelis constant. They both depend on the rate constants of the reaction, and v_{max} also depends on the concentration of the enzyme. The details of the derivation are beyond us for now, but you will see in the following exercises how this model behaves for different values of A.

2.5 Computational projects

In this project you will plot the graphs of the models you computed in the computational assignment at the end of Chapter 1. Here are the models again:

1. Heart pumping with a constant rate, where V_{tot} is the total volume of blood pumped by the heart over time, V_s is the stroke volume, R is the heart rate and t is time:

$$V_{tot} = V_s R t$$

2. Bacterial population that doubles every hour, where P is the current population, P_0 is the initial population, and t is the elapsed time:

$$P = P_0 2^t$$

3. The rate of an enzyme-catalyzed reaction, where v is the rate of the reaction, v_{max} is the maximum reaction rate, K_M is the

Michaelis constant, and A is concentration of the substrate:

$$v = \frac{v_{max} A}{K_M + A}$$

Computational tasks

1. Plot the dependence of total blood pumped by the heart on the elapsed time, based on the heart pumping model. To do this, set time to be a vector from 0 to 80 minutes with step 1 minute and calculate a vector of values of volume using your script from computational task 1.1 (in Chapter 1) and keeping the same values of stroke volume and heart rate. Plot the volume vs time using the type line with black line color, and label your axes. Then change the heart rate to 120 beats per minute and overlay that plot on top of the previous one using the lines() function with red line color. Based on the plots, describe what kind of function this model is and the difference between the two graphs.

2. Plot the dependence of bacterial population on time, based on the bacterial population model. To do this, set time to be a vector from 0 to 3 days with step of 1 hour, and calculate a vector of values of population using your script from computational task 1.2 and keeping the same initial population. Plot the population vs time using the type line with black line color, and label your axes. Then change the initial population to 2 million and overlay that plot on top of the previous one using the lines() function with red line color. Based on the plots, describe what kind of function this model is and the difference between the two graphs.

3. Plot the dependence of the reaction rate on the substrate concentration, based on the Michaelis-Menten model. To do this, set substrate concentration to be a vector from 0 to 100 mM with step of 1 mM, and calculate a vector of values of reaction rate using your script from computational task 1.3 and keeping the same values for all the other variables. Plot the reaction rate vs concentration using the type line with black

2.5. COMPUTATIONAL PROJECTS

line color, and label your axes. Based on the plot, at what value of concentration is the reaction rate equal to half the maximum rate v_{max}? Change the Michaelis constant K_M to be 60 mM, overlay that plot on top of the previous one using the `lines()` function with red line color, and report at what concentration the reaction rate is half of v_{max}. How is this value influenced by K_M?

Chapter 3

Describing data sets

> *Get your facts first, and then you can distort them as much as you please.*
> —Rudyard Kipling, *An Interview with Mark Twain*

Science begins with experimental measurements, which are then verified by reproducing the results. But no experimental result is perfectly reproducible, because all are subject to random noise, whether it is caused by unpredictable processes or is due to measurement error. Describing collections of numbers with noise is the first step to understanding the biological systems that are being measured. In this chapter you will learn to do the following:

1. calculate the mean and the median of a data set,

2. calculate the variance and the standard deviation of a data set,

3. produce histograms and interpret them, and

4. use R to plot and analyze data sets.

3.1 Mutations and their rates

All Earth-based life forms receive an inheritance from their parent(s): a string of deoxyribonucleic acids (*DNA*) called the genetic sequence, or *genome*, of an individual. The information to produce all the necessary components to build and maintain the organism is encoded in the sequence of the four different *nucleotides*: adenine, thymine, guanine, and cytosine (abbreviated as A, T, G, and C, respectively). Different parts of the genome play different roles; some discrete chunks called *genes* contain the instructions to build *proteins*, the workhorses of biology. To make a protein from a gene, the information is *transcribed* from DNA into *messenger ribonucleic acid* (*mRNA*), which is then *translated* into a string of *amino acids*, which constitute the protein. The *genetic code* determines the translation, using three nucleic acids in DNA and RNA to represent a single amino acid in a protein. Thus, a sequence of DNA always results in a specific sequence of amino acids, which determine the structure and function of the protein.

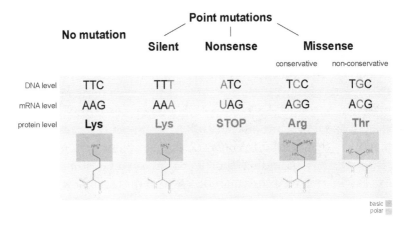

Figure 3.1: Different types of substitution point mutations are distinguished by their effects on the gene products. Image by Jonsta247 in public domain via Wikimedia Commons.

The above processes involve copying and transferring information. As we know from experience, copying information inevitably means introducing errors. This is particularly important when passing information from parent to offspring, because then an entire organism has to develop and live based on a faulty blueprint. Changes introduced in the genome of an organism are called *mutations*, and they can be caused either by errors in copying DNA when making a

3.1. MUTATIONS AND THEIR RATES

new cell (replication) or through damage to DNA through physical means (e.g., ionizing radiation) or chemical mechanisms (e.g., exogenous molecules that react with DNA). The simplest mutations involve a single nucleotide and are called *point mutations*. A nucleotide may be deleted, an extra nucleotide inserted, or a new one substituted instead: the three different types of substitution mutations are shown in Figure 3.1. Large-scale mutations may involve whole chunks of the genome that are cut out and pasted in a different location, or copied and inserted in another position, but they are typically much more rare than point mutations.

Mutations can have different effects on the mutant organism, although acquisition of super-powers has not been observed. Usually, point mutations have either little observable effect or a negative effect on the health of the mutant. A classic example is *sickle-cell disease*, in which the molecules of the protein hemoglobin, responsible for carrying oxygen in the blood from the lungs to the tissues, tends to stick together and clump, resulting in sickle-shaped red blood cells. The disease is caused by a single substitution mutation in the gene that codes for one of the two components of hemoglobin, called β-globin. The substitution of a single nucleotide in the DNA sequence changes one amino acid in the protein from glutamate to valine, which causes the proteins to aggregate. This *missense* mutation (see Figure 3.1) is carried by a fraction of the human population, and those who inherit the mutations from both parents develop the painful and sometimes deadly disease. Such mutations that are present in some but not all of a population are called *polymorphisms*, to distinguish them from mutations that occurred in evolutionary lineages and differentiate species from one another.

One of the central questions of evolutionary biology is: how frequently do mutations occur? Since mutations are generally undesirable, most living things have developed ways to minimize the frequency of errors in copying DNA, and to repair DNA damage. But even though mutations are rare, they occur spontaneously in all organisms, because molecular processes such as copying a DNA molecule are subject to random noise arising from thermal motion. Mutations are fundamentally a random process, and we need to use *descriptive statistics* to analyze data with inherent randomness.

3.2 Describing data sets

3.2.1 central value of a data set

A data set is a collection of measurements. These measurements can come from many kinds of sources and can represent all sorts of quantities. One big distinction is between numerical and categorical data sets. *Numerical* data sets contain numbers, either integers or real numbers. Some examples: number of individuals in a population, length, blood pressure, concentration. *Categorical* data sets may contain numbers, symbols, or words, limited to a discrete (usually small) number of values. The word "categorical" is used because this kind of data corresponds to categories or states of the subject of the experiment. Some examples: genomic classification of an individual on the basis of one locus (e.g., wild type or mutant), the state of an ion channel (open or closed), the stage of a cell in the cell cycle.

A data set contains more than one measurement; the number of measurements is called the size of the data set and is usually denoted by the letter n. To describe a data set numerically, one can use numbers called *statistics* (not to be confused with the branch of science of the same name). The most common statistics aim to describe the central value of the data set to represent a typical measurement. If you order all the measurements from highest to lowest and then take the the middle value, you have found the *median* (if there is an even number of values, take the average between the middle two). Precisely half the data values are less than the median and the other half are greater, so the median represents the true "middle" value of the measurement. Note that the median can be calculated either for numerical or categorial data, as long as the categories can be ordered in some fashion.

The value that occurs most frequently in the data set is called its *mode*. For some data sets, particularly those which are symmetric, the mode coincides with the mean (see next paragraph) and the median, but for many others it is distinct. The mode is the most visual of the three statistics, as it can be picked out from the histogram plot of a data set (which is described in section 3.2.3) as the value corresponding to the maximum frequency. The mode can also be used for both categorical and numerical data.

3.2. DESCRIBING DATA SETS

The *average*, or *mean* of a data set is the sum of all the values divided by the number of values. It is also called the *expected value*, because it allows us to simply predict the sum of a large number of measurements with a given mean, by multiplying the mean by the number. The mean can be calculated only for a numerical data set, since we cannot add non-numerical values. The definition of the mean of a data set X is the same as for the average and is usually indicated with a bar over the letter for the measured data variable:

$$\bar{X} = \frac{1}{n} \sum_{i=1}^{n} x_i \qquad (3.1)$$

The mean, unlike the median, is not the middle value of the data set; instead it represents the "center of mass" of the measured values (Whitlock and Schluter 2008). Another way of thinking of the mean is as a weighted sum of the values in the data set. The weights represent the frequency of occurrence of each numeric value in the data set, which we will further discuss in the section 3.2.3.

The mean is the most frequently used statistic, but it is not always interpreted correctly. Very commonly the mean is reported as the most representative value of a data set, but that is often misleading. There are at least two situations in which the mean can be tricky: (1) data sets with a small number of discrete values; and (2) data sets with *outliers*, or isolated numbers very far from the mean.

Examples of misleading means. Mean values for data sets with a few quantities are not the typical value, such as in the number of children born in a year per individual, also known as the birth rate. The birth rate per year in 2013 for both the United States and Russia is 1.3% per person, but you will have to look for a long time to find any individual who gave birth to 1.3% of a child. While this point may be obvious, it is often overlooked when interpreting mean values.

Outliers are another source of trouble for means. For example, a single individual (let's call him or her B.G.) with a wealth of $50 billion moves into a town of 1000 households with average wealth of $100,000. Although none of the original residents' assets have changed, the mean wealth of the town improves dramatically, as you can calculate in one of the exercises at the end of the chapter. One can cite the improved per capita wealth in the town as evidence of economic growth, but that is obviously misleading. In cases with such dramatic outliers, the median is the more informative representation of a typical value of the data set.

Math exercises:

For the (small) data sets given below, calculate the mean and the median (by hand or using a calculator) and compare the two measures of the center.

Exercise 3.2.1. Data set of the population of the city of Chicago (in millions) in the past four census years (2010, 2000, 1990, 1980): 2.7 2.9 2.8 3.0.

Exercise 3.2.2. Data set of the numbers of the fish known as the blacknose dace (*Rhinichthys atratulus*) collected in six different streams in the Rock Creek watershed in Maryland: 76 102 12 55 93 98.

Exercise 3.2.3. Data set of tuberculosis incidence rates (per 100,000 people) in the five largest metropolitan areas in the United States in 2012: 5.2 6.6 3.2 5.5 4.5.

Exercise 3.2.4. Consider the hypothetical town with 1000 households with mean and median wealth of $100,000 and one person with assets of $50 billion. Calculate the mean value of the combined data set, and compare it to the new median value.

Exercise 3.2.5. Data set of ages of mothers at birth for five individuals: 19 20 22 32 39.

Exercise 3.2.6. Data set of ages of fathers at birth for five individuals: 22 23 25 36 40.

Exercise 3.2.7. Data set of the number of new mutations found on maternal chromosomes for five individuals: 9 10 11 26 15.

Exercise 3.2.8. Data set of the number of new mutations found on paternal chromosomes for five individuals: 39, 43, 51, 53, 91.

3.2. DESCRIBING DATA SETS

Thinking problem:

Problem 3.2.1. Suppose you'd like to add a new observation to a data set (e.g., the sixth largest metropolitan area (Philadelphia) to the tuberculosis incidence data set, which is 3.0). Calculate the mean of the 6-value data set, without using the 5 values in the original data set, but only using the mean of the 5-value data set and the new value. Generalize this to calculating the sample mean for any n-value data set, given the mean of the $n-1$ values, plus one new value.

3.2.2 spread of a data set

The center of a data set is obviously important, but so is its spread around the center. Sometimes the spread is caused by noise or error, for example, in a data set of repeated measurements of the same variable under the same conditions. At other times the variance is due to real changes in the system or to inherent randomness of the system, and the size of the spread-as well as the shape of the histogram—are important for understanding the mechanism. The simplest way to describe the spread of a numerical data set is to look at the difference between the maximum and minimum values, called the *range*. However, it is obviously influenced by outliers, since the extreme values are used. To describe the typical spread, we need to use all the values in the data set and see how far each one is from the center, as measured by the mean.

There is a problem with the naive approach: if we just add up all the differences of data values from the mean, the positives will cancel the negatives, and we'll get an artificially low spread. One way to correct this is to take the absolute value of the differences before adding them up. However, for somewhat deep mathematical reasons, the standard measure of spread uses not absolute values, but squares of the differences, and then divides that sum not by the number of data points n but by $n-1$. This is called the *variance* of a data set, and it is calculated by the formula (Whitlock and Schluter 2008):

$$Var(X) = \frac{1}{n-1} \sum_{i=1}^{n} (\bar{X} - x_i)^2 \qquad (3.2)$$

The variance is a sum of square differences, so its dimension is the square of the dimensions of the measurements in X. To obtain a measure of the spread comparable to the values of X, we take the square root of variance and call it the *standard deviation* of the data set X:

$$\sigma(X) = \sqrt{\frac{1}{n-1} \sum_{i=1}^{n} (\bar{X} - x_i)^2} \qquad (3.3)$$

Just as the mean is a weighted average of all of the values in the data set, the variance is a weighted average of all the squared deviations of the data from the mean.

Math exercises:

For the (small) data sets given below, calculate the range, variance, and standard deviation (by hand or using a calculator). Compare the range and the standard deviation for each case: Which one is larger? By how much?

Exercise 3.2.9. Data set of the population of the city of Chicago (in millions) in the past four census years (2010, 2000, 1990, 1980): 2.7 2.9 2.8 3.0.

Exercise 3.2.10. Data set of the numbers of the fish known as the blacknose dace (*Rhinichthys atratulus*) collected in six different streams in the Rock Creek watershed in Maryland: 76 102 12 55 93 98.

Exercise 3.2.11. Data set of tuberculosis incidence rates (per 100,000 people) in the five largest metropolitan areas in the United States in 2012: 5.2 6.6 3.2 5.5 4.5.

Exercise 3.2.12. Data set of ages of mothers at birth for five individuals: 19 20 22 32 39.

Exercise 3.2.13. Data set of ages of fathers at birth for five individuals: 22 23 25 36 40.

Exercise 3.2.14. Data set of the number of new mutations found on maternal chromosomes for five individuals: 9 10 11 26 15.

3.2. DESCRIBING DATA SETS

Exercise 3.2.15. Data set of the number of new mutations found on paternal chromosomes for five individuals: 39, 43, 51, 53, 91.

Thinking problem:

Problem 3.2.2. Suppose that a data set has a fixed range (e.g., all values have to lie between 0 and 1). What is the greatest possible standard deviation for any data set within the range? Hint: Think about how to place the points as far from the mean as possible. How do the data sets given above relate to your prediction?

3.2.3 graphical representation of data sets

Data sets can be presented visually to indicate the frequency of different values. This can be done in a number of ways, depending on the kind of data set. For a data set with only a few values (e.g., a categorical data set), a good way to represent it is with a *pie chart*. Each category is represented by a slice of the pie with the area of the same share of the pie as the fraction of the data set in the category. There is some evidence, however, that pie charts can be misleading to the eye, so R does not recommend using them.

For a numerical data set it is useful to plot the frequencies of a range of values, which is called a *histogram*. Its independent axis has the values of the data variable, and the dependent axis has the frequency of those values. If the data set consists of real numbers that range across an interval, that interval is divided into subintervals (usually of equal size), called *bins*, and the number of measurements in each bin is indicated on the y-axis. To be visually informative, there should be a reasonable number (usually no more than a few dozen, although it varies) of bins. The most frequent/common measurements are represented as the highest bars or points on the histogram. Histograms can show either the counts of measurements in each bin, or the fraction of the total number of measurements in each bin. The only difference between those two kinds of histogram is the scale of the y-axis, and, confusingly, both can be called frequencies.

A histogram of the measured lengths of the bacterium *Bacillus subtilis* is shown in Figure 3.2. The data set was measured in

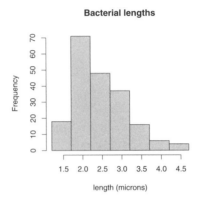

Figure 3.2: Length of bacteria *Bacillus subtilis* measured under the microscope as discrete values with step of 0.5 μm data from Watkins (n.d.).

increments of a half-micron, with numbers varying between 1.5 and 4.5 microns. The histogram shows that the most common measurement (the mode) is 2 μm. Adding up all frequencies in the histogram tells us that there are approximately 200 values in the data set. This allows us to find the median value by counting the frequencies of the first few bins until we get to 100 (the median point), which resides in the bin for 2.5 μm. It is a little bit more difficult to estimate the mean, but it should be clear that the center of mass of the histogram is also near 2.5 (it is actually 2.49). Finally, the hardest task is estimating the spread of the data set, such as the standard deviation, based on the histogram. The range of the data set is $4.5 - 1.5 = 3$, so we know for sure that it is less than 1.5. The histogram shows that the deviations from the mean value of 2.5 range from 2 (rarely) to 0.5 (most prevalent). This should give you an idea that the weighted average of the deviations is less than 1. Indeed, the correct standard deviation is about 0.67.

There are different ways of plotting data sets that have more than one variable. For instance, a data set measured over time is called a *time series*. If the values are plotted with the corresponding times on the x-axis, then it is called a *time plot*. This is useful to

3.3. WORKING WITH DATA IN R

show the changes of the values of a variable over time. If the data set doesn't undergo any significant changes over time, it makes more sense to represent it as a pie chart or histogram. More generally, one may plot two variables measured together on a single plot, which is called a *scatterplot*. We will explore such plots and the relationships between two measured variables in Chapter 8.

Math exercises:

Answer the following questions, based on the histograms in Figure 3.3 (mutation data) and Figure 3.4 (heart rate data).

Exercise 3.2.16. How many people in the mutation data have fathers either younger than 20 or older than 40? How many have more than 80 new mutations?

Exercise 3.2.17. Estimate the mode, median, and mean of the two variables in the mutation data set.

Exercise 3.2.18. State the range of each data set, and estimate the standard deviation of the two variables in the mutation data set.

Exercise 3.2.19. How many people in the heart rate data plot (Figure 3.4) have heart rates greater than 80 bpm? How many have body temperature less that 97° F?

Exercise 3.2.20. Estimate the mode, median, and mean of the two variables in the heart rate data set.

Exercise 3.2.21. State the range of each data set, and estimate the standard deviation of the two variables in the heart rate data set.

3.3 Working with data in R

One way to input data into R is to read in a text file, where several variables are stored in columns. For instance, the file HR_temp.txt contains three variables: body temperature (in degrees Fahrenheit),

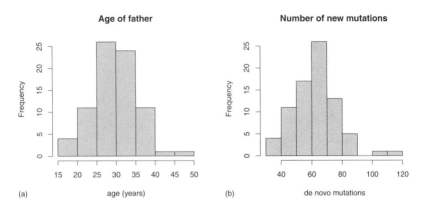

Figure 3.3: Histograms of (a) paternal ages and (b) the number of new mutations from 73 families; data from Kong et al. (2012).

```
data <- read.table("data/HR_temp.txt", header = TRUE)
hist(data$HR, col = "gray", main = "Heart rate data",
    xlab = "heart rate (bpm)")
hist(data$Temp, col = "gray",
    main = "Body temperature data",
    xlab = "temperature (F)")
```

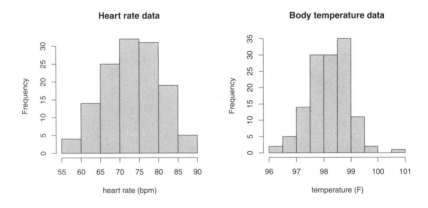

Figure 3.4: Histograms of heart rates and body temperatures.

3.3. WORKING WITH DATA IN R

sex (1 for male, 2 for female), and heart rate (in beats per minute). The values for the variables are arranged in columns, and the first row of the file contains the names of the variables (Temp, Sex, and HR, respectively). The R command `read.table()` reads this file and and puts it into a *data frame* called data. The three variables are stored inside the data frame and can be accessed by appending the dollar sign and variable name to the data frame, so data$HR refers to only the heart rates, and data$Temp refers to the body temperatures.

The code example in Figure 3.4 shows how to read in a data file and plot the histograms for multiple data sets. The script below demonstrates the basic commands for describing a data set in R. The `hist()` function makes a plot of the histogram of the results; `median()` and `mean()` compute the median and the mean, respectively; `var()` and `sd()` compute the variance and the standard deviation, respectively. Finally, a useful command called `summary()` prints out some basic descriptive statistics for each variable in the data frame.

```
mean(data$HR)
```
```
## [1] 73.76154
```
```
var(data$HR)
```
```
## [1] 49.87293
```
```
sd(data$HR)
```
```
## [1] 7.062077
```
```
summary(data)
```
```
##       Temp            Sex            HR
##  Min.   : 96.30   Min.   :1.0   Min.   :57.00
##  1st Qu.: 97.80   1st Qu.:1.0   1st Qu.:69.00
##  Median : 98.30   Median :1.5   Median :74.00
##  Mean   : 98.25   Mean   :1.5   Mean   :73.76
##  3rd Qu.: 98.70   3rd Qu.:2.0   3rd Qu.:79.00
##  Max.   :100.80   Max.   :2.0   Max.   :89.00
```

3.4 Computational projects

Before starting the first project, measure a data set of your heart rate values, which you will analyze using R. You can do this by downloading the free app called Heart Rate (for either iPhone or Android phones) and measure your heart rate 10 times (while resting). If you don't have a smartphone, you can measure your heart rate manually, or borrow someone else's phone. Record the heart rate values in a text file as a column of numbers, which you can use in R Studio by creating a new text file from the File menu. When entering the data, you can use the first row to make a one-word label for the column, which, after R reads in the data, will become the name of the variable in the data frame. Do not enter anything other than a single number for the data values (e.g., do not enter 75 bpm, just 75).

3.4.1 data description

For this project you will use the data set of your heart rates and a data set of one of your classmate's heart rates. Make sure you save all of the data files and your .Rmd report file to the same folder on your computer, and set the working directory in R Studio to that folder. Use the command read.table() with the option header=TRUE to read in the the data files into different data frames with descriptive names. Once you've loaded the data, you can access the individual variables and calculate their statistics.

Computational tasks

1. Read in your data file of heart rates. Report its mean, range, and standard deviation, and compare the range and the standard deviation. Then stand up and walk around the lab and measure your heart rate again. Report the new measurement and whether it is within the range of the data set you measured at rest.

2. Read in the data file your classmate's heart rates and report the statistics (mean, range, standard deviation). Do the ranges of the two data sets overlap? Combine the two data

3.4. COMPUTATIONAL PROJECTS

sets into one (either by putting together the text files or by using the command c() introduced in Chapter 2), and report the statistics of the combined data set. Plot its histogram and comment on whether you can distinguish the two original data sets on the plot.

3.4.2 plotting data and fitting by eye

In this section you will read in three different data files and assign each data frame a descriptive name (e.g., KaiC, Insulin, and Kong). The data files can be downloaded from the textbook website: https://dkon.uchicago.edu/page/quantifying-life. The contents of the data files are described as follows:

1. Read the data on number of mutations and the ages of the father and the mother (Kong et al., 2012) found in the file kong_mutation_data.txt, consisting of variables PatAge and MatAge (paternal and maternal ages, respectively, in years) and the variable Mutations (number of de novo mutations).

2. Read in the data for concentration of insulin in blood plasma of a human after administration of glucose found in the file Insulin_data.txt, consisting of variables Conc (concentration of insulin in micrograms/mL) and the variable Time (in minutes after peak concentration).

3. Read in the data for the phosphorylated circadian clock protein called KaiC (Akiyama et al. 2008) found in the file KaiC_data.txt, consisting of variables Amount (in arbitrary units) and Time (hours).

In the following tasks, you will plot the dependent variable as a function of the independent variable, decide whether it resembles a linear function $f(x) = mx + b$ or an exponential function $f(x) = ae^{kx} + b$, and then find reasonable parameters for the function that fit the data reasonably well. After plotting the data as circles, use the **curve()** command to overlay a plot of a function of the dependent variable. For example, to plot the data and overlay a straight line with slope m and intercept b, execute the following script:

```
plot(data$time, data$population)
curve(x*m+b,min(data$time),max(data$time),add=TRUE)
```

Your task is to choose the parameters (m and b for a line or a, b, and k for an exponential curve) that produce a plot that does a decent job of resembling the data. To find the parameters that best fit the data set, choose two of them so that (a) the value of the given function at zero matches the value of the data set at zero and (b) the value the function approaches at infinity matches the value that the data set approaches in the long run and then (c) try different values of the remaining parameter to see which one makes the function resemble the data between 0 and infinity.

Computational tasks

1. Make a plot of Insulin$Conc vs Insulin$Time using the default type (circles only). Decide whether it resembles a linear or exponential function, and estimate two of the parameters from the behavior at small and large values of the independent variable as described above, until you have only one free parameter left, and then try different values to see which one produces a curve that best fits the data. Report: (a) which function you chose; (b) what parameters you estimated from the behavior at extremes; and (c) which value of the free parameter you tried was the best match. You should get the graph of the function to at least resemble the data plot.

2. Make a plot of KaiC$Amount vs KaiC$Time using the default type (circles only). Decide whether it resembles a linear or exponential function, and estimate the parameters from limiting data values as described above, until you have only one free parameter left, and then try different values to see which one produces a curve that best fits the data. Report: (a) what function you chose; (b) what parameters you estimated from limiting behavior; and (c) which value of the free parameter you tried was the best match. You should get the graph of the function to at least resemble the data plot.

3.4. COMPUTATIONAL PROJECTS

3. Make a plot of Kong$Mutations vs Kong$PatAge using the default type (circles only). Decide whether it resembles a linear or exponential function, and estimate the parameters from limiting data values as described above, until you have only one free parameter left, and then try different values to see which one produces a curve that best fits the data. Report: (a) what function you chose; (b) what parameters you estimated from limiting behavior; and (c) which value of the free parameter you tried was the best match. You should get the graph of the function to at least resemble the data plot.

4. Make a plot of Kong$Mutations vs Kong$MatAge using the default type (circles only). Decide whether it resembles a linear or exponential function, and estimate the parameters from limiting data values as described above, until you have only one free parameter left, and then try different values to see which one produces a curve that best fits the data. Report: (a) what function you chose; (b) what parameters you estimated from limiting behavior; and (c) which value of the free parameter you tried was the best match. You should get the graph of the function to at least resemble the data plot.

Chapter 4

Random variables and distributions

> *What is there then that can be taken as true? Perhaps only this one thing, that nothing at all is certain.*
> —René Descartes

Mathematical models can be divided into *deterministic* and *stochastic* models. Deterministic models assume that the future can be perfectly predicted based on complete information of the past. Stochastic models instead assume that even perfect knowledge of the past does not allow one to predict the future with certainty.

Stochastic models may not sound very promising: after all, we want to make predictions, and randomness says that predictions are impossible! However, the word "random" in mathematics doesn't mean "completely unpredictable" or "without rules," as it does in common usage. It means that we can make probabilistic predictions (e.g., compute what fraction of molecules will diffuse from one place to another, or what fraction of genes mutate in one generation)—we just can't make a definite prediction for each individual molecule or gene. Biological processes are so complex and are subject to so much environmental noise that stochastic models are absolutely essential for understanding of many living systems. Here is what you will learn to do in this chapter:

1. define probability in terms of outcomes and events,
2. recognize random variables and their distributions,
3. compute means and variances of distributions,
4. use the binomial distribution to model strings of binary trials,
5. generate random numbers in R, and
6. use loops to repeat calculations in R.

4.1 Probability distributions

4.1.1 axioms of probability

In this section we develop the terminology used in the mathematical study of randomness called probability. This begins with a *random experiment*, which is a broad term that can describe any natural or theoretical process whose outcome cannot be predicted with certainty. The number of outcomes may be *discrete* (can be counted by integers) or *continuous* (corresponding to real numbers). We will stick with experiments that have discrete outcomes in this chapter, but many important experiments produce continuous outcomes. The first step for studying a random process is to describe all the outcomes it can produce:

Definition 4.1. The collection of all possible outcomes of an experiment is called its *sample space* Ω. An *event* is a subset of the sample space, which means an event may contain one or more experimental outcomes.

Example. You can ask a person two questions: "how tall are you?" (and classify them either as short or tall) and "do you like tea?" (yes or no), and you've performed a random experiment. The randomness comes not from the answers (assuming the person doesn't randomly lie) but from the selection of the respondent. We will discuss randomly selecting a sample from a population in Chapter 5. This random experiment has four outcomes: tall person who likes tea, tall person who does not like tea, short person who likes

4.1. PROBABILITY DISTRIBUTIONS

Figure 4.1: The sample space of all people with two events: tall people and those who like tea.

tea, and short person who does not like tea. This sample space and events is illustrated in Figure 4.1 with a Venn diagram, which uses geometric shapes as representations of events as subsets of the entire sample space. These outcomes can be grouped into events by one of the responses: for example, tall person (A) or person who doesn't like tea ($-B$).

Example. A random experiment with two outcomes, called a *Bernoulli trial* (named after Jacob Bernoulli, one of the founders of mathematical probability), can describe a variety of situations: a coin toss (heads or tails), a competition with two outcomes (win or loss), the allele of a gene (normal or mutant). The sample space for a single Bernoulli trial consists of just two outcomes: $\{H, T\}$ (for a coin toss). If the experiment is performed repeatedly, the sample space gets more complicated. For two Bernoulli trials there are four different outcomes $\{HH, HT, TH, TT\}$. One can define different events for this sample space: the event of getting two heads in two tosses contains one outcome, $\{HH\}$; the event of getting a single head contains two, $\{TH, HT\}$.

To describe the composition of a sample space, we need to define the word *probability* (Feller 1968). While it is familiar to everyone from everyday usage, it is difficult to define without using other similar words, such as likelihood or plausibility, which also need to be defined. We can agree that something with a high probability happens often, while something with a low frequency is seldom observed. Thus, one of the accepted definitions of probability of an outcome of a random experiment is the *fraction of experiments with this outcome out of many repeated experiments.* This is a conceptual definition rather than a practical one, because to measure this fraction exactly one needs to repeat the experiments many times. This definition is called the *frequentist* view of probability, in contrast with the *Bayesian* approach that we will investigate in Chapter 7. The difference between the two approaches is partly philosophical and partly practical: a frequentist supposes that a random experiment can be repeated many times, while a Bayesian seeks to use information from a limited number of experiments.

Once we have defined the probability of an outcome, we can calculate the probability of a collection of outcomes according to rules that ensure the results are self-consistent. These rules can be summarized in the following definition.

Definition 4.2. The *probability* $P(A)$ of an event A in a sample space Ω is a number between 0 and 1, which obeys the following rules, called the *axioms of probability*:

1. $P(\Omega) = 1$

2. $P(\emptyset) = 0$

3. $P(A \cup B) = P(A) + P(B) - P(A \cap B)$

Let us define some notation for sets: $A \cup B$ is called the *union* of two sets, which contains all elements of A and B; $A \cap B$ is called the *intersection* of two sets, which contains all elements that are in both A and B; and $\emptyset$ denotes the empty set. Any event A has its *complement*, denoted $-A$, which contains all elements of Ω that

4.1. PROBABILITY DISTRIBUTIONS

are *not* in A. These operations on sets are illustrated in Figure 4.2. Applying them to the sample space and events in Figure 4.1, the union of the two sets $A \cup B$ are all people who are either tall or like tea, the intersection of the two sets $A \cap B$ are all the tall people who like tea, and the intersection of the first set with the complement of the second $A \cup -B$ are all tall people who do not like tea.

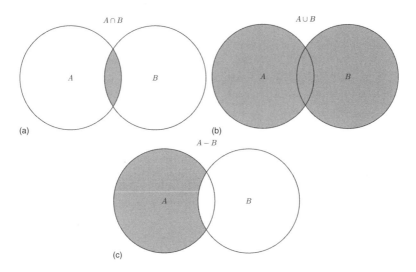

Figure 4.2: Operations on sets with shaded areas denoting the result: (a) intersection of two sets, (b) union of two sets, (c) the intersection of a set with the complement of another.

The first two axioms are based on intuition: the probability of *some* outcome from the sample space occurring is 1, and the probability of *nothing* in the sample space occurring is 0. The intuition behind axiom three is less transparent, but it can be see in a Venn diagram of two subsets A and B of the larger set Ω, as in Figure 4.1. Compare the size of the union of A and B and the sum of the sizes of sets A and B separately, and you will see that the intersection $A \cap B$ occurs in both A and B, but it is only counted once in the union. This is why it needs to be subtracted from the sum of $P(A)$ and $P(B)$.

There are several useful rules that immediately follow from the axioms. First, if two events are mutually exclusive—meaning their intersection is empty ($A \cap B = \emptyset$)—then the probability of either of them happening is the sum of their respective probabilities: $P(A \cup B) = P(A) + P(B)$ (from axiom 3). Further, since an event A and its complement $-A$ are mutually exclusive, their union is the entire sample space Ω: $P(A)+P(-A) = P(A\cup -A) = P(\Omega) = 1$; therefore $P(A) = 1 - P(-A)$.

Example. Assume one is using a fair coin, so the probability of a single head and a single tail is 1/2. Assuming the result of the first toss does not affect the next one, the probability of getting two heads in a row is 1/4, because after the coin comes up heads once, half of the time it will come up heads again. In fact, the probability of getting any particular sequence of two coin toss results is 1/4. Here are some examples of what we can calculate:

- The probability of getting one head of out of two tosses is $1 - 1/4 - 1/4 = 1/2$ (by the complement rule).

- The probability of not getting two heads is $1 - 1/4 = 3/4$ (by the complement rule).

- The probability of getting either zero, one, or two heads is 1 (by axiom 1).

- The probability of getting three heads is 0 (since this event is not in the sample space).

Example. Suppose one is testing people for a mutation that has the probability (prevalence) of 0.2 in the population, so for each person, there are two possible outcomes: normal or mutant. The probability of drawing two mutants in a row is $0.2 \times 0.2 = 0.04$ by the same argument as above; the probability of drawing two normal people is $0.8 \times 0.8 = 0.64$. Based on this, we can calculate the following.

- The probability of one mutant out of two people is $1 - 0.04 - 0.64 = 0.32$ (by the complement rule).

- The probability of not having two mutants is $1 - 0.04 = 0.96$ (by the complement rule).

- The probability of either zero, one, or two mutants is 1 (by axiom 1).

4.1. PROBABILITY DISTRIBUTIONS

- The probability of getting three mutants is 0 (since this event is not in the sample space).

4.1.2 random variables

The outcomes of experiments can be expressed in numbers or words, but we generally need numbers to report and analyze results. One can describe this mathematically as a function (recall its definition form section 2.2) that assigns numbers to random outcomes (Feller 1968). In practice, a random variable describes the measurement that one makes to describe the outcomes of a random experiment.

Definition 4.3. A *random variable* is a function that takes outcomes from the sample space and assigns numbers to them.

Example. Define the random variable to be the number of heads out of two coin tosses. This random variable will return numbers 0, 1, or 2, corresponding to different events. The random variable of the number of mutants out of two people (assuming there are only two outcomes, mutant and normal) has the same set of values.

A random variable has a set of possible values, and each of those values may come up more or less frequently in a random experiment. The frequency of each measurement corresponds to the probability of the outcomes in the sample space that produce that particular value of the random variable. One can describe the behavior of the random variable in terms of the collection of the probabilities of its outcomes.

Definition 4.4. The probability of a random variable X taking some value x, written as $P(X = x)$ (but usually simplified to $P(x)$) is the probability of the event corresponding to the value x of the random variable. This function $P(x)$ is called the *probability distribution* of the random variable X.

One important property of every discrete probability distribution function is that all its values have to add up to 1:

$$\sum_{i=1}^{n} P(x_i) = 1$$

The graph of a probability distribution function lies above zero, because all probabilities are between 0 and 1. The graph of a probability distribution is similar to a histogram: it represents the frequency of occurrence of each value of the random variable. A histogram is an approximation of the true probability distribution based on a sample, and for a large sample size, the histogram approaches the graph of the probability distribution function.

Example. Assuming that each coin toss has probability 1/2 of resulting in heads, the probability distribution function for the number of heads out of two coin tosses is $P(0) = 1/4$, $P(1) = 1/2$, and $P(2) = 1/4$ (as we computed in the example in the previous section). Note that the probabilities add up to 1, as they should.

Example. For the random variable of the number of mutants out of two people, for a mutation prevalence of 0.2, the probability distribution function is $P(0) = 0.64$, $P(1) = 0.32$, and $P(2) = 0.04$ (as we computed in the example in the previous section). Note that the probabilities add up to 1, as they should.

4.1.3 expectation (mean) of random variables

Definition 4.5. The *expected value* or *mean* of a discrete random variable X with probability distribution $P(X)$ is defined as

$$E(X) = \mu_X = \sum_{i=1}^{n} x_i P(x_i)$$

This sum is over all values $\{x_i\}$ that the random variable X can take, multiplied by the probability of the random variable taking that value (meaning the probability of the event in sample space that corresponds to that value). This corresponds to the definition of the mean of a data set given in section 3.2, if you consider $P(x_i)$ to be the number of times x_i occurs divided by the number of total measurements n. As in the case of the histogram and the probability distribution function, the mean of a sample for a large sample size n approaches the mean of the random variable, as I discuss in more detail in Chapter 5. Sometimes we will use the more concise $\mu_X = E(X)$ to represent the mean (expected) value. Here are some basic properties of the expectation.

4.1. PROBABILITY DISTRIBUTIONS

1. The expectation of a random variable that is a constant (c) is equal to c, since the probability of c is 1: $E(c) = cP(c) = c$.

2. The expectation of a constant multiple of a random variable is

$$E(cX) = \sum_i cx_i P(x_i) = c \sum_i x_i P(x_i) = c\mu_X$$

3. The expectation of a sum of two random variables is the sum of their expectations. This is a more complicated argument, so let us break it down. First, all possible values of the random variable $X+Y$ come from going through the possible values of X and Y, and for each value of x_i and y_j, we need to multiply the sum by the probability of x_i and y_j:

$$E(X+Y) = \sum_i \sum_j (x_i + y_j) P(y_j) P(x_i)$$

Because x_i does not depend on the value of j, we can take it out of the sum over j and then break up the sum. Then, because adding up $P(y_j)$ over all values of j is 1, and because the second term contains the mean of Y, we get

$$E(X+Y) = \sum_i \left[x_i P(x_i) \sum_j P(y_j) + P(x_i) \sum_j y_j P(y_j) \right] =$$
$$= \sum_i [x_i P(x_i) + P(x_i) \mu_Y]$$

Finally, we break up the sum over i, and since μ_Y does not depend on i, we can do the same simplification as above:

$$E(X+Y) = \sum_i x_i P(x_i) + \mu_Y \sum_i P(x_i) = \mu_X + \mu_Y$$

Example. The expected value of the number of heads out of two coin tosses can be calculated using the probability distribution function we found in the example with fair coin in section 4.1.2:

$$E(X) = 0 \times P(0) + 1 \times P(1) + 2 \times P(2) = 0 + 1/2 + 2 \times 1/4 = 1$$

The expected number of heads out of 2 is 1, if each head comes up with probability 1/2, which I think you will find intuitive.

Example. The expected value of the number of mutants out of two people can be calculated using the probability distribution function we found in the example in subsection 4.1.2:

$$E(X) = 0 \times P(0) + 1 \times P(1) + 2 \times P(2) = 0 + 1 \times 0.32 + 2 \times 0.04 = 0.4$$

The expected number of mutants in a sample of two people is 0.4, which may seem a bit strange. Recall that mean or expected values do not have to coincide with values that are possible, as we discussed in section 3.2, but are instead a weighted average of values, according to their frequencies or probabilities.

4.1.4 variance of random variables

Knowledge of the expected value says nothing about how the random variable actually *varies*: expectation does not distinguish between a constant random variable and one that can deviate far from the mean. To quantify this variation, you might be tempted to compute the mean differences from the mean value, but it does not work:

$$E(X - \mu_X) = \sum_i (x_i - \mu_x) P(x_i) = \sum_i x_i P(x_i) - \mu_x \sum_i P(x_i)$$
$$= \mu_x - \mu_x = 0$$

The problem is, if we add up all the differences from the mean, the positive ones end up canceling the negative ones. This is why it makes sense to square the differences and add them up.

Definition 4.6. The *variance* of a discrete random variable X with probability distribution $P(x)$ is

$$Var(X) = E((X - \mu_X)^2) = \sum_{i=1}^{n} (x_i - \mu_x)^2 P(x_i)$$

4.1. PROBABILITY DISTRIBUTIONS

One highly useful property of the variance is
$$Var(X) = \sum_i (x_i^2 - 2x_i\mu_x + \mu_x^2)P(x_i) =$$
$$= \sum_i x_i^2 P(x_i) - 2\mu_x \sum_i x_i P(x_i) + \mu_x^2 \sum_i P(x_i) =$$
$$= E(X^2) - \mu_x^2$$

So variance can be calculated as the difference between the expectation of the variable squared and the squared expectation. Note that the variance is given in units of the variable squared, so to measure the spread of the variable in the same units, we take the square root of the variance and call it the *standard deviation*: $\sigma_x = \sqrt{Var(X)}$.

Example. The variance of the number of heads out of two coin tosses can be calculated using its probability distribution function and the expected value of 1 we found in the example in section 4.1.3:
$$Var(X) = (0-1)^2 \times P(0) + (1-1)^2 \times P(1) + (2-1)^2 \times P(2)$$
$$= 1/4 + 0 + 1/4 = 1/2$$

Since the variance is 1/2, the standard deviation, or the expected distance from the mean value, is $\sigma = \sqrt{1/2}$.

Example. The variance of the number of mutants out of two people can be calculated using its probability distribution function and the expected value of 0.4 we found in the example in section 4.1.3:
$$E(X) = (0-0.4)^2 \times P(0) + (1-0.4)^2 \times P(1) + (2-0.4)^2 \times P(2)$$
$$= 0.4^2 \times 0.64 + 0.6^2 \times 0.32 + 1.6^2 \times 0.04 = 0.32$$

Since the variance is 0.32, the standard deviation, or the expected distance from the mean value, is $\sigma = \sqrt{0.32}$.

Math exercises:

Calculate the expected values and variances of the following probability distributions, where the possible values of the random variable are in curly brackets, and the probability of each value is indicated as $P(x)$.

Exercise 4.1.1. $X = \{0, 1\}$ and $P(0) = 0.1, P(1) = 0.9$.

Exercise 4.1.2. $X = \{1, 2, 3\}$ and $P(1) = P(2) = P(3) = 1/3$.

Exercise 4.1.3. $X = \{10, 15, 100\}$ and $P(10) = 0.5, P(15) = 0.3, P(100) = 0.2$.

Exercise 4.1.4. $X = \{0, 1, 2, 3, 4\}$ and $P(0) = 1/8, P(1) = P(2) = P(3) = 1/4, P(4) = 1/8$.

Exercise 4.1.5. $X = \{-1.5, -0.4, 0.3, 0.9\}$ and $P(-1.5) = 0.4$, $P(-0.4) = 0.2, P(0.3) = 0.35, P(0.9) = 0.05$.

4.2 Examples of distributions

4.2.1 uniform distribution

Perhaps the simplest random variable (besides a constant, which is not really random) is the *uniform random variable*, for which every outcome has equal probability. The distribution of a fair coin is uniform with two values, H or T, or 0 and 1, each with probability 1/2. More generally, a discrete uniform random variable has n outcomes and each one has probability $1/n$. This is what people often mean when they use the word random—an experiment where each outcome is equally likely. Probability distribution functions for two uniform random variables (on integers between 0 and 3 and between -6 and 3) are plotted in Figure 4.3.

It is straightforward to calculate the expectation and variance of a uniform random variable.

$$E(X) = \sum_{i=1}^{n} x_i P(x_i) = \frac{1}{n} \sum_{i=1}^{n} x_i \qquad (4.1)$$

So the expected value is the mean of all the values of the uniform random variable.

Example. For the random variable on integers from 1 to n, the mean value is the average of the maximum and minimum values (using the fact that $\sum_{i=1}^{n} i = n(n+1)/2$):

$$\frac{n^2 + n}{2n} = \frac{n+1}{2}$$

4.2. EXAMPLES OF DISTRIBUTIONS

Generalizing, for a random variable on integers between a and b, the expectation is $E(x) = \frac{a+b}{2}$. We can also write down the expression for the variance:

$$Var(X) = E(X^2) - \mu_X^2 = \frac{1}{n}\sum_{i=1}^{n} x_i^2 - \frac{1}{n^2}\left(\sum_{i=1}^{n} x_i\right)^2$$

Example. For the uniform distribution on n integers between 1 and n ($x_i = i$, for $i = 1, ..., n$), we can calculate the variance using the formula for the sum of squares: $\sum_{i=1}^{n} i^2 = n(n+1)(2n+1)/6$:

$$\frac{(n+1)(2n+1)n}{6n} - \frac{n^4 + 2n^3 + n^2}{4n^2} = \frac{2n^2 + 3n + 1}{6} - \frac{n^2 + 2n + 1}{4}$$

$$= \frac{n^2}{12} - \frac{1}{12} \qquad (4.2)$$

Generalizing, for a uniform random variable on integers between a and b, the variance is

$$Var(X) = \frac{(b - a + 1)^2 - 1}{12}.$$

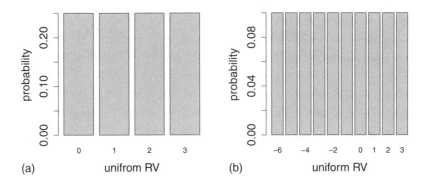

Figure 4.3: Two uniform random distributions with integer values with different ranges.

4.2.2 binomial distribution

We already introduced binary or Bernoulli trials in section 4.1. Assume that the two values of the random variable X are 0 and 1, with probability $1-p$ and p, respectively. Then we can calculate the expectation and variance of a single Bernoulli trial:

$$E(X) = 0 \times (1-p) + 1 \times p = p$$

$$Var(X) = E(X^2) - \mu_X^2 = 0^2 \times (1-p) + 1^2 \times p - p^2 = p(1-p)$$

The first result is likely intuitive, but the second deserves a comment. Note that depending on the probability of 1, the variance—or the spread in outcomes of a Bernoulli trial—is different. The highest variance occurs when $p = 1/2$ (equal probability of 0 or 1), but when p approaches 0 or 1, the variance approaches 0. Thus, as the probability approaches 0 or 1, the random variable approaches a constant (either always 1 or 0); hence, no variance.

One can extend this scenario and ask what happens in a string of Bernoulli trials, for instance, in a string of 10 coin tosses, or in testing 20 people for a mutation. The mathematical problem is to calculate the probability distribution of the number of success out of many trials. This is known as the binomial random variable, which is defined as the sum of n independent, identical Bernoulli random variables.

Definition 4.7. Given n independent Bernoulli trials with the same probability of outcomes, the *binomial random variable* is defined as (Feller 1968)

$$B = \sum_{i=1}^{n} X_i$$

where X_i is the random variable from the ith Bernoulli trial, which takes values of 1 and 0.

In this definition I use the term "independence" without defining it properly, which will be done in section 6.2. Intuitively, independence between two Bernoulli trials (e.g., coin tosses) means that the outcome of one trial does not change the probability of the outcome the other trial. This amounts to the assumption that the probability of an outcome followed by another one is the product of the separate probabilities of the two outcomes. For example, if the two

4.2. EXAMPLES OF DISTRIBUTIONS

outcomes are wins and losses, then $P(\{WL\}) = P(W)P(L)$. This will be used below in the calculation of the variance of the binomial random variable.

To find the probability distribution of the binomial random variable, we need to define the event "k wins out of n trials." Consider the case of 4 trials. It is easy to find the event of 4 wins, as it is comprised only of the outcome $\{WWWW\}$. Then, $P(4) = p^4$, based on the independence assumption. The event of winning 3 times consists of four strings: $\{LWWW, WLWW, WWLW, WWWL\}$ so the probability of obtaining 3 wins is the sum of the four probabilities, each equal to $p^3(1-p)$ from the independence assumption above, so $P(3) = 4p^3(1-p)$. The event of winning 2 times is even more cumbersome, and consists of six strings: $\{LLWW, WLLW, WWLL, WLWL, LWLW, LWWL\}$, so $P(2) = 6p^2(1-p)^2$ by the same reasoning.

Now imagine doing this to calculate 50 wins out of 100 trials. The counting gets ugly very fast. We need a general formula to help us count the number of ways of winning k times out of n trials. We denote this number $\binom{n}{k}$, also known as "n choose k," because it corresponds to the number of ways of choosing k distinct objects out of n, without regard to order. The connection is as follows: let us label each trial from 1 to n. Then to construct a string with k wins, we need to specify which trials resulted in a win (the rest are of course losses). It does not matter in which order those wins are selected—it still results in the same string.

Thus, we can count as follows: there are n possibilities for choosing the number of the first win, then $n-1$ possibilities for choosing the number of the second win, and so forth. Finally when choosing the kth win, there are $n-k+1$ possibilities (note that $k \leq n$, and if $n = k$ only one option is left for the last choice.) Thus, the total number of such selections is: $n(n-1)\cdots(n-k+1) = n!/(n-k)!$

But note that we overcounted, because we considered different strings of wins depending on the order in which a win was selected, even if the resulting strings are the same (e.g., $n = 4$ and $k = 4$ gives us 4!, although there is only one string of 4 wins out of 4). To correct for the overcounting, we need to divide by the total number of ways of selecting the same string of k wins out of n. This is

number of ways of rearranging k wins, or $k!$. Thus, the number we seek is
$$\binom{n}{k} = \frac{n!}{k!(n-k)!}$$

We can now calculate the general probability of winning k times out of n trials. First, each string of k wins and $n-k$ losses has the probability $p^k(1-p)^{n-k}$. Since we now know that the number of such strings is C_k^n, the probability is

$$P(k \text{ wins in } n \text{ trials}) = P(B = k) = \binom{n}{k} p^k (1-p)^{n-k} \qquad (4.3)$$

This is the probability distribution of the binomial random variable B.

The binomial random variable has much simpler formulas for the mean and the variance. First, we know that the mean of a sum of random variables is the sum of the means, and the binomial random variable is a sum of n Bernoulli random variables X. Let us say X takes only the values 0 and 1 with probabilities q and p, respectively, so we compute the expected value of B:

$$E(B) = E\left[\sum_{i=1}^{n} X\right] = \sum_{i=1}^{n} E(X) = \sum_{i=1}^{n} p = np \qquad (4.4)$$

Intuitively, this says that the expected number of heads/successes is the product of the probability of 1 head/success and the number of trials (e.g., if the probability of success is 0.3, then the expected number of successes out of 100 is 30).

Now let us calculate the variance, for which in general the same property is not true. It turns out that under the assumption of independence, the variance of a sum of random variables is the sum of their respective variances. Here is the derivation:

$$Var(X+Y) = E\left[(X+Y) - (\mu_X + \mu_Y)\right]^2 =$$
$$= E[(X - \mu_X)^2 + (Y - \mu_Y)^2 - 2(X - \mu_X)(Y - \mu_Y)] =$$
$$= E(X - \mu_X)^2 + E(Y - \mu_Y)^2 - 2E[(X - \mu_X)(Y - \mu_Y)] =$$
$$= Var(X) + Var(Y) - 2E[(X - \mu_X)(Y - \mu_Y)]$$

4.2. EXAMPLES OF DISTRIBUTIONS

The last term is known as the *covariance* of the two random variables X and Y. Here I use the property of independent random variables stated above, that $P(x,y) = P(x)P(y)$, to show that the covariance of two independent random variables is 0:

$$E((X - \mu_X)(Y - \mu_Y)) = \sum_i \sum_j (x_i - \mu_X)(y_j - \mu_Y) P(x_i, y_j) =$$

$$= \sum_i (x_i - \mu_X) P(x_i) \sum_j (y_j - \mu_Y) P(y_j)$$

We saw in section 4.1 that the expected value of deviations from the mean is zero, which gives

$$E((X - \mu_X)(Y - \mu_Y)) = E(X - \mu_X) E(Y - \mu_Y) = 0$$

This demonstrates that for independent variables, the variance of their sum is the sum of the variances, and therefore for the binomial random variable we have

$$Var(B) = Var\left[\sum_{i=1}^n X\right] = \sum_{i=1}^n Var(X)$$

$$= \sum_{i=1}^n p(1-p) = np(1-p) \tag{4.5}$$

For any given number of Bernoulli trials, the variance has a quadratic dependence on probability of success p: if $p = 1$ or $p = 0$, corresponding to all successes, or all failures, respectively, then the variance is 0, since there is no spread in the outcome. For a fair coin with $p = 1/2$, the variance is highest. This can be seen in the plots of binomial random variables for $n = 2$, $n = 5$, and $n = 50$, shown in Figures 4.4, 4.5, and 4.6, respectively.

Math exercises:

Calculate the means and variances based on the plotted distributions using definitions 4.5 and 4.6 and compare your calculations to equations 4.1 and 4.2 (for uniform random variables) and equations 4.4 and 4.5 (for binomial random variables).

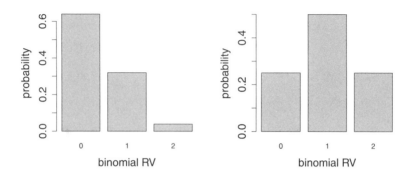

Figure 4.4: The binomial distribution for $n = 2$ and $p = 0.2$ and $p = 0.5$. (You should be able to tell which one is which!)

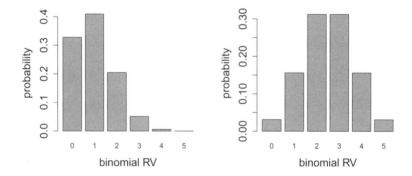

Figure 4.5: The binomial distribution for $n = 5$ and $p = 0.2$ and $p = 0.5$. (You should be able to tell which one is which!)

Exercise 4.2.1. Calculate the mean and the variance for the two uniform distributions plotted in Figure 4.3.

Exercise 4.2.2. Calculate the mean and the variance for the two binomial distributions plotted in Figure 4.4.

Exercise 4.2.3. Calculate the mean and the variance for the two binomial distributions plotted Figure 4.5.

Exercise 4.2.4. Estimate (approximately) the mean and the variance for the two binomial distributions plotted in Figure 4.6.

4.3. TESTING FOR MUTANTS

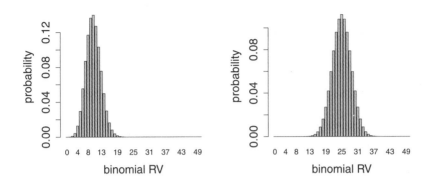

Figure 4.6: The binomial distribution for $n = 50$ and $p = 0.2$ and $p = 0.5$. (You should be able to tell which one is which!)

4.3 Testing for mutants

Suppose that you're screening people for a particular genetic abnormality. It is known from prior experience that about 5% of this population carries this mutation. You run your tests on a group of 20 people, and the results indicate that 3 of them are carriers. Clearly, this is higher than you expected: 3/20 is 15%, or 3 times higher than the estimate. One of your colleagues exclaims, "What are the odds of that?"

To answer this question, you must start by stating your assumptions. First, the people tested must be chosen from the same population, so we can assume a priori each had probability 5% of being a carrier. Second, the people must be selected without bias; that is, selection of one must be unlinked or independent of selection of the others. As a counterexample, if your selection included an entire biological family, that would be a biased selection—it may be that the whole family has the mutation, or maybe they don't, but either way, probability is no longer determined on a person-by-person basis. If these assumptions are made, then you can calculate the probability of making a selection of 20 people that includes 3 carriers of the mutation, using the binomial distribution.

The formula for the binomial distribution in equation 4.3 provides the answer for any given number of mutants. For example, the probability of 3 people out of 20 being carriers for the mutation is

$$P(3 \text{ out of } 20; p = 0.05) = \binom{20}{3} \times 0.05^3 \times 0.985^{17} =$$
$$= 1140 \times 0.05^3 \times 0.985^{17} \approx 0.0596$$

You may want to ask a different question: what is the probability that there are at least 3 mutants in the sample of 20 people? The most efficient way to calculate this is to answer the complementary question first: what is the probability that there are fewer than 3 mutants out of 20 people? This corresponds to three values of the random variable: 0, 1, or 2. We can calculate the total probability by adding up the three separate probabilities, since they represent nonoverlapping events (one can't have 1 and 2 mutants in a sample simultaneously):

$$P(B < 3; p = 0.05) = P(B = 0) + P(B = 1) + P(B = 2) =$$

$$= \binom{20}{2} \times 0.05^2 \times 0.95^{18} + \binom{20}{1} \times 0.05^1 \times 0.95^{19}$$
$$+ \binom{20}{0} \times 0.05^0 \times 0.95^{20} \approx 0.925$$

The answer to the original question is found by taking the complementary probability $1 - 0.925 = 0.075$. Thus the probability of finding at least 3 mutants in a sample of 20 with individual probability 0.05 is approximately 0.075. The answer is so close to the probability of having exactly 3 mutants because the probability of finding more than 3 mutants is very low.

4.4 Random numbers and iteration in R

4.4.1 random numbers

Simulating randomness with a computer is not a simple task. Randomness is contrary to the nature of a computer, which is designed

4.4. RANDOM NUMBERS AND ITERATION IN R

to perform operations exactly. However, there are algorithms that produce a string of numbers that are for all intents and purposes random: there is no obvious connection between one number and the next, and the values don't form any pattern. Such algorithms are called *random number generators*, although to be more precise they produce *pseudo-random* numbers. The reason is that they actually produce a perfectly predictable string of numbers, which eventually repeats itself, but with a humongous period. One can even produce the same random number, or the same string of random numbers, by specifying the *seed* for the random number generator. This is very useful if one wants to reproduce the results of a code that uses random numbers.

Of course, random variable are not all the same—they have different distributions. R has several functions for producing random numbers from different distributions. For example, to produce random numbers from a set of values with a uniform probability distribution, use the function `sample()`. The following command produces a random integer between 1 and 20. Repeating the same command produces a new random number, which (most likely) will not be the same as the first. The first input argument (1:20) is the vector of values from which to draw the random number, and the second is the size of the sample:

```
sample(1:20, 1)
```

```
## [1] 15
```

```
sample(1:20, 1)
```

```
## [1] 3
```

To generate 10 randomly chosen integers between 1 and 20, see the following two commands, which differ in setting the value of the option "replace." The first command doesn't specify the value for replace, and by default it is set to FALSE, so the command draws numbers without replacing them (meaning that all the numbers in the sample are unique). In the second command, "replace" is set to TRUE, so the numbers that were selected can be chosen again. In both cases, repeatedly running the command results in a different set of randomly chosen numbers, which you should investigate by copying the commands into R and running them yourself.

```
sample(1:20, 10)
```

```
##  [1]  7  2 11 18 16  4 17  8 20 14
```

```
sample(1:20, 10, replace = TRUE)
```

```
##  [1]  8 12  9 19  3  5 20  3  6  6
```

If you need to generate a random number from the binomial distribution, R has you covered. The function is rbinom(s, n, p), and it requires three input values: s is the number of observations (sample size), n is the number of binary trials in one observation, and p is the probability of success in one binary trial. The following two commands generate a single random number, the number of successes out of 20 trials with probability of success 0.2 and 0.6:

```
rbinom(1,20,0.2)
## [1] 5
rbinom(1,20,0.6)
## [1] 11
```

To generate an entire sample of random numbers, change the first input parameter to 10. As you'd expect, the samples of 10 observations are (most likely) noticeably different: when the probability p is 0.2, the number of successes tend to be less than 6, while for probability 0.6, the numbers are usually greater than 10.

```
rbinom(10, 20, 0.2)
```

```
##  [1] 6 6 5 5 4 4 7 2 3 3
```

```
rbinom(10, 20, 0.6)
```

```
##  [1] 11 13 10 12 12 12 11 13 11 14
```

4.4. RANDOM NUMBERS AND ITERATION IN R

Notice that the range of possible values of this random variable is between 0 and 20, but unlike the uniform random numbers produced with the `sample()` function, the probability of obtaining different numbers are different and depend on the parameter p. Calculation and plotting of the binomial distribution function can be accomplished with the function `dbinom(x,n,p)`, where x is the value of the random variable (between 0 and n), n is the number of trials, and p is the probability of success. This is illustrated in the script in Figure 4.7, which first uses the `dbinom()` function to compute the probability of obtaining exactly 2 successes out 20 trials when the probability of success in one trial is 0.2. Then the script calculates the probabilities of all of the possible values of the random variable by substituting the vector of these values (e.g., 0 to 20) instead of the number 1, generating the probability distribution vector. This vector is plotted vs the values of the random variable using the `barplot()` function, producing an aesthetically pleasing plot of the binomial distribution. The script plots two binomial probability distributions, both with $n = 20$, the first with $p = 0.2$ and the second with $p = 0.6$. Notice also the use of the axis labels in barplot using the same options we saw before (`xlab` and `ylab`) and the use of the `main` option to produce a title above each plot.

4.4.2 for loops

Often times one has to write a script to perform the same (or similar) task many times. Like all programming languages, R has commands that call for repetition of the same commands multiple times. These structures are called *loops*. In this section you will learn about the *for loop*, which looks like:

```
for (i in 1:10) {
    body of the loop
}
```

The loop starts with the special command `for` followed by parentheses and the special command `(i in 1:10)`. This tells the loop how many times to repeat its contents, called the *body of the loop*. The body consists of any number of commands inside the curly

```
n <- 20
p <- 0.2
dbinom(1, n, p)

## [1] 0.05764608

values.vec <- 0:n
prob.dist <- dbinom(values.vec, n, p)
barplot(prob.dist, names.arg = values.vec,
    xlab = "binomial RV",
    ylab = "probability", main = "binom dist with n=20
        and p=0.2")
p <- 0.6
prob.dist <- dbinom(values.vec, n, p)
barplot(prob.dist, names.arg = values.vec,
    xlab = "binomial RV",
    ylab = "probability", main = "binom dist with n=20
        and p=0.6")
```

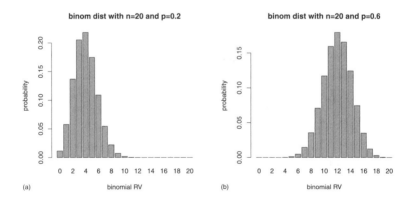

Figure 4.7: Two binomial probability distributions (plotted using the R script above).

4.4. RANDOM NUMBERS AND ITERATION IN R

braces, so the commands inside the braces are executed repeatedly (as many times as you want), and the commands outside the braces (either before or after the loop) are only executed once. In the above example, the loop above instructs R to execute the body 10 times, while increasing the counter variable i from 1 to 10, inclusive. The variable i can be used in the loop itself, where it will change in value from 1 to 10. For example, the following script adds up all the integers from 1 to 10 and prints them out:

```
total <- 0 for (i
in 1:10) {
    total <- total + i
} print(total)
```

```
## [1] 55
```

You can use a for loop to create an entire vector variable, for instance, by keeping a running total while adding up the integers from 0 to 10. To do that, you will have to *pre-allocate* the vector before you start assigning values to its elements. Here's how to preallocate a vector with 10 placeholders called NA (Not Available) instead of numbers, and then use it to store the running total of the 10 integers:

```
running.total <- rep(NA, 10)
running.total[1] <- 1
for (i in 2:10) {
    running.total[i] <- running.total[i - 1] + i
}
print(running.total)
```

```
##  [1]  1  3  6 10 15 21 28 36 45 55
```

Let us return to generating random numbers from a uniform distribution. Using the for loop enables us to generate multiple samples, as can be seen in the following script. Here, we generate 30 samples of uniform random numbers between 1 and 20 and save their mean values in a vector. This vector has to be initialized (which we will do by filling it with zeros) before the beginning of the for loop:

```
num.samples <- 30
mean.vec <- rep(0, num.samples)
for (i in 1:num.samples) {
    data <- sample(1:20, 10)
    mean.vec[i] <- mean(data)
}
print(mean.vec)

##  [1]  9.8  9.0 10.6  9.8  9.6 11.3 14.1  8.5 10.4
## [10] 10.1 10.6 10.6 10.3 11.2 10.6 10.8 11.0  8.9
## [19] 13.3 12.3  9.2  8.9 10.2 11.8 10.2  9.6 11.3
## [28] 10.4 11.0 11.5
```

Notice that I used the same variable name `data` to save the sample of 10 random numbers every iteration, and then reused the same vector to store the next sample. We didn't need access to data outside the for loop, so it makes sense to "recycle" a variable multiple times. In contrast, the means of the different random samples needed to be saved, so each one was placed in a different element of the `mean.vec` vector. The counter variable `i`, which increases with every iteration of the for loop, provides a convenient means of specifying the index of the vector.

4.5 Computational projects

In these projects you will use random number generators to simulate different biological experiments. The first simulation is based on the uniform distribution, and the second on the binomial distribution.

4.5.1 uniform distribution

In this project we investigate the uniform distribution, which is what people typically mean when they say something is "random." Suppose that there are three different species of eel in the same coral

4.5. COMPUTATIONAL PROJECTS

reef, and assume that they are equally prevalent, and therefore each one has equal probability of getting caught.

1. Generate a random sample of 3 eels, whose species is indicated by integers between 1 and 3. Is the frequency of each species in agreement with their expected probabilities?

2. Generate 10 random samples of 3 eels, and report how many of those have the expected probabilities. Hint: Use a for loop.

3. Generate a random sample of 30 eels, and make a histogram of the frequencies of the three species. How do they compare to the expected probabilities? Run the simulation several times, and report how much the histogram changes each time.

4. Generate a random sample of 300 eels, and make a histogram of the frequencies of the three species. How do they compare to the expected probabilities? Explain how larger sample size affects the histogram of observations and how closely it matches the true distribution of eels in the population. Run the simulation several times, and report how the histogram changes each time.

4.5.2 binomial distribution

Let us return to the questions posed in section 4.3. Suppose that a gene exists in a population in two versions: normal and mutant, and the mutants make up 3% of the population. Use the binomial random number generator rbinom() to generate the number of mutants in of one or more samples of a specified size, and use dbinom() to calculate the probability of different outcomes.

1. Generate a random number of mutants out a sample of 20 people and report it. Suppose you had only this number for estimating the fraction of mutants in the population. How far off would you be from the truth?

2. Generate 100 random numbers of mutants from samples of 20 people (no need to use a for loop), and report them by plotting

a histogram. How many of those samples have 3 mutants? Calculate the actual probability of having 3 mutants out of 20 and compare it to your observation.

3. Generate 100 random numbers of mutants from samples of 200 people and report them by plotting a histogram. How many of those samples have 30 mutants? Calculate the actual probability of having 30 mutants out of 200, and compare it to your observation.

Chapter 5

Estimation from a random sample

Pragmatism!? Is that all you have to offer?
—Tom Stoppard, *Rosencrantz and Guildenstern Are Dead*

One of the most common tasks in any experimental science is to measure the value of a quantity by performing repeated experiments. A single experiment is not sufficient, because there are always random factors making any single observation imperfect, such as natural variations of the random variable or experimental errors. For example, the number of new mutations (those not present in either parent) in an offspring is a random variable, so each individual may have a different number of them. A set of experimental measurements is called the *sample*. In this chapter you will learn to do the following:

1. understand the sample mean as a random variable,

2. calculate the standard error for a sample mean,

3. calculate the confidence interval for the mean,

4. understand the meaning of the parameters of the normal distribution, and

5. simulate sampling of a random variable using R.

5.1 Law of Large Numbers

5.1.1 sample mean

The goal in this chapter is to calculate the best estimate of the true mean value of a random variable. To do this, we need to repeat the relevant experiment several times and take the mean of the data set: if one measurement is too high while another one is too low, their average is much closer to the true value. One important consideration here is that the errors cannot be systematically *biased*. If your experiment always overestimates or underestimates the true value, then averaging the measurements will result in a biased estimate. Another consideration is that the errors cannot depend on one another: if one measurement is above average, it should have no effect on the probability of other measurements being above average.

In statistical parlance, a perfectly representative, unbiased sample is called a *simple random sample*. Speaking precisely, for a sample with n observations (size n), generating such a sample requires that all subsets of the population of the same size have equal probability of being selected. In practice, this is not easy to verify. Sampling bias occurs if the way a sample is selected results in some participants being systematically overrepresented and other underrepresented. With a biased sample, any conclusions from a study are suspect, and the mathematics is no longer applicable.

In statistics, it is customary to distinguish between the *population* and the *sample*. The term "population" doesn't always refer to a collection of people or other living creatures; instead it may refer to all possible outcomes of an experiment—essentially, the entire sample space that introduced in Chapter 4. Out of this ocean of possibilities, an experimentalist fishes out a subset called the sample, and uses it to describe the whole population. It should be clear that the sample needs to be representative of the whole population, if this endeavor is to have any hope of success.

As mentioned in section 4.5, the most common estimation involves the mean. Here we need to distinguish two concepts: the *true mean* and the *sample mean*. The true mean refers to the quantity we are trying to measure (estimate), and it is considered the mean of a probability distribution of a random variable across the entire

5.1. LAW OF LARGE NUMBERS

population. Thus the true mean is a *parameter* that we are trying to estimate.

The sample mean $\bar{X}$ is the average of a sample of experimental measurements, and it is a random variable with its own probability distribution. Suppose you've collected a sample of size n and measured its mean $\bar{X}_1$. If you collect another sample of the same size, its mean $\bar{X}_2$ will be (most likely) a different number. Repeating this process many times will result in different values of the sample mean, all dancing around the true mean. Provided that the data collection was unbiased each time, the variation is due to a combination of variance of the random variable in the population and the *sampling error*. The different sample means result from selecting samples that may randomly contain more numbers that are higher than the true mean than are lower (or the opposite).

In this and the following sections, I will describe the variation in the sample mean. It is a remarkable fact that we can describe this distribution, called the *sampling distribution of the mean*, in general, and connect it to the true distribution of the variable in the population. The first result describes the mean of the sampling distribution (Watkins n.d.):

Theorem 5.1. (Law of Large Numbers, part 1) For a sample of n independent measurements with the same distribution with mean μ, the sample mean approaches μ as n becomes large. More formally, if $X_1, X_2, ..., X_n$ are independent, identically distributed random variables with mean μ, then

$$\lim_{n \to \infty} \frac{X_1 + X_2 + \cdots + X_n}{n} = \lim_{n \to \infty} \bar{X} = \mu$$

This result is called the *Law of Large Numbers*, and it is commonsensical: as the size of an unbiased random sample increases, its sample mean approaches the true mean. If we are dealing with a finite population, then of course if the sample includes the entire population, its mean will be the true mean.

One consequence of this is that the best estimator for the true mean is the sample mean; in fact it is what is called an *unbiased estimator*, provided the sampling process is unbiased. However, this is of only of limited use to someone trying to estimate a quantity, because the theorem doesn't say how large a sample size needs to be for its mean to be reasonably close to the true mean. To address this practical question, let us consider the variance of the sampling distribution.

5.1.2 sample size and standard error

As mentioned above, repeated samples from the same populations have varying sample means, and we want to describe the variance. This variation depends on the sample size, as suggested by the Law of Large Numbers: for large sample sizes, the sample mean approaches the true mean. The intuitive fact is that the larger the sample size, the less variation there will be in the sampling distribution.

Theorem 5.2. (Law of Large Numbers, part 2) For a sample of n independent measurements with the same distribution with variance σ^2, the variance of the sampling distribution for a sufficiently large n approaches the variance divided by the sample size:

$$Var(\bar{X}) = \frac{\sigma^2}{n}$$

In words, for a sample of n independent measurements, the variance of the sample mean is inversely proportional to the sample size. Here I sketch a calculation to justify (if not prove) the theorem. Suppose that the distribution of experimental measurements has expected value of μ and variance σ^2. Then for a sample of n independent measurements $\{X_i\}$, the variance of the sample mean is

$$Var(\bar{X}) = Var\left(\frac{1}{n}\sum_{i=1}^{n} X_i\right) = \frac{1}{n^2}\sum_{i=1}^{n} Var(X_i) = \frac{1}{n^2}n\sigma^2 = \frac{\sigma^2}{n}$$

The critical step in that calculation is the second equal sign, where the variance passes inside the sum. This operation is valid because, as we saw in section 4.2, the variance of a sum of independent random variables is equal to the sum of the variances.

It is often useful to deal with standard deviations to describe the spread of a distribution. Since the standard deviation is the square root of variance, we have the following definition.

5.2. CENTRAL LIMIT THEOREM

Definition 5.1. The standard deviation of the mean of a sample of n independent, identically distributed random variables with standard deviation σ is called the *standard error* of the sample mean. For large sample size n, according to the Law of Large Numbers, it approaches $s = \sigma/\sqrt{n}$.

The standard error represents the spread in estimation of the true mean based on the data set of size n. Thus, increasing the sample size by a factor of 100 will lead to a reduction in the spread of the sample mean by a factor of 10. For example, if we obtain a sample of size 1000 instead of 10, the sample mean will have much less volatility.

5.2 Central Limit Theorem

5.2.1 normal distribution

In section 5.1 we calculated the mean and standard deviation of sampling distributions for a large sample size. There is an even more remarkable fact about sampling distributions: they all look the same! Regardless of the distribution of the random variable being sampled, the plot of the distribution of sample means approaches the famous bell-shaped curve called the *normal distribution*. This is one of the most fundamental and useful results in all of mathematics. It is called the Central Limit Theorem.

The theorem is much more subtle in both its statement and implications than the Law of Large Numbers, so I need to spell out a few preliminaries. First, the sample mean for a large sample size can take on a whole range of values, so we can think of it as a *continuous random variable.* Say you're sampling from the uniform distribution of integers between 1 and 10. If your sample size is 2, the mean may be either an integer (e.g., 5 if your sample is 6 and 4) or a fraction with denominator 2 (e.g., 11/2 if your sample is 1 and 10). As the sample size grows larger, the denominator of the sample mean increases, and the possible means are smeared out more densely between the values of the discrete distribution (e.g., for sample size 100, you may observe a sample mean of 3.62). As the sample sizes become larger, it is convenient to stipulate that the

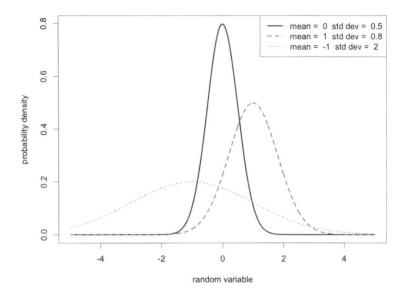

Figure 5.1: Normal probability densities with different parameter values. The mean μ determines the center of the distribution, and the standard deviation σ controls the width.

range of values of the sample mean is continuous, that is, it contains all real numbers between the maximum and the minimum values of the original distribution.

This brings us to the magnificent, famous, and indispensable normal (or Gaussian) distribution. More correctly, it is the probability *density* function of the normal random variable, which has the range from negative infinity to infinity, that is, any real number. Here is the mathematical form of its density function $\rho(x)$, with parameters μ and σ representing the mean and the standard deviation, respectively (Whitlock and Schluter 2008):

$$\rho(x) = \sqrt{\frac{1}{2\pi\sigma^2}} e^{-\frac{(x-\mu)^2}{2\sigma^2}} \quad (5.1)$$

The function of the random variable x is plotted in Figure 5.1, for different values of the mean and the standard deviation. The shape of the distribution is the famous bell-shaped curve, and the mean indicates the position of the peak on the x-axis. The standard

5.2. CENTRAL LIMIT THEOREM

deviation parameter is responsible for the width of the distribution: the larger σ is, the broader and more spread out the distribution will be. Those are the only parameters in the normal distribution; everything else about it remains the same.

This probability density function differs from the discrete distribution functions we saw in Chapter 4 in one key way: although you can plug in a particular value of the normal random variable, the number it returns is not the probability of that particular value. Properly speaking, the probability of any particular value of the normal random variable, like 3.2, is indistinguishable from zero, because there are infinitely many such values, and if they all had nonzero probabilities, the total probability of the distribution would be infinite. In fact, the total probability of the normal random variable must be 1, as dictated by the axioms of probability given in definition 4.2.

Instead, to extract probability from a density function, it must be integrated. The integral is the equivalent of a sum when you need to add up numbers from a continuous range. So instead of calculating the probability of one value, for continuous random variables we can calculate the probability of a range of values by taking the integral of the density function $\rho(x)$ on that range. For example, the probability of the normal random variable with $\mu = 1$ and standard deviation $\sigma = 0.8$ being in the range between 0 and 1 is:

$$P(0 < x < 1) = \sqrt{\frac{1}{2\pi 0.8^2}} \int_0^1 e^{-\frac{(x-1)^2}{2 \times 0.8^2}} dx$$

These integrals may look intimidating, and rightly so: there is no method for solving them the way you may have done in a calculus course. But although we cannot write down a closed formula, we can still find the answer *numerically*, that is, as a number. In the old days, one consulted a table in the back of the probability or statistics textbook that contained the values of these integrals. We can do this much more efficiently using R, as demonstrated in sections 5.4 and 5.5.

Now that we have a nodding acquaintance with the normal distribution, here is its most wonderful property: the means of samples, for large sample sizes, are distributed normally (Watkins n.d.).

Theorem 5.3. (Central Limit Theorem) The distribution of the mean of a sample of n independent, identically distributed random variables with mean μ and standard deviation σ approaches for large n the *normal distribution* with mean μ and standard deviation $\sigma/\sqrt{n}$. If the above conditions are met, the probability that a sample mean will fall between values a and b is given by the integral of the probability density function:

$$P(a < \bar{X} < b) = \sqrt{\frac{n}{2\pi\sigma^2}} \int_a^b e^{-\frac{(x-\mu)^2}{2\sigma^2/n}} dx$$

This theorem gifts us the ability to predict how widely the means of a sample can vary around the true mean. If you know the true mean μ, the true standard deviation σ, and the sample size n is large enough (ignoring for a second what that means exactly), then the probability of the sample mean being off by 0.1 from the true mean in either direction is

$$P(\mu - 0.1 < \bar{X} < \mu + 0.1) = \sqrt{\frac{n}{2\pi\sigma^2}} \int_{\mu-0.1}^{\mu+0.1} e^{-\frac{(x-\mu)^2}{2\sigma^2/n}} dx$$

To evaluate this integral (numerically), we need to know the values of the three parameters μ, σ, and n. In practice, only the sample size n is known, while μ and σ are parameters of the true distribution and can be only estimated. The next section is devoted to sorting out the statistical details.

5.2.2 confidence intervals

The Central Limit Theorem is a purely mathematical result, but it is used in a wide range of practical applications, from public opinion polling to medical risk assessments. In this section we journey from the abstract land of probability to the data-driven domain of statistics on a quest for the correct estimation of means. When reporting any experimental measurement, it is mandatory to include *error bars* around the mean of a data set to indicate a range of plausible values of the estimated quantity, usually in the form of $\bar{X} \pm \epsilon$. The meaning of these error bars varies: sometimes the standard deviation of the measurements (σ_X) is used, at other times the standard

5.2. CENTRAL LIMIT THEOREM

error is used, but the correct way to report uncertainty in estimation is to calculate the confidence interval (Whitlock and Schluter 2008).

Definition 5.2. A *confidence interval* is a range of values calculated from a data set to estimate the true value of a quantity (e.g., the mean). The associated *confidence level* α (between 0 and 1) is the likelihood that the confidence interval contains the true value.

The meaning of confidence intervals is pretty subtle. This is in large part because of the word "likelihood" in the definition, which in everyday language is interchangeable with "probability," but as mathematical terms they are not. Here is the wrong way to think about them: a confidence interval for a mean at α level does **not** mean that the true mean has probability α of being in that interval. The true mean is not a random variable, it is a parameter of the probability distribution that we assume exists. Instead, it means that if sampling were repeated many times, fraction α of the resulting confidence intervals would contain the true mean.

You can see from this definition that there is more than one confidence interval one can report from a single data set, because its size depends on the confidence level. At first glance, you might be tempted to make α as large as possible—after all, you'd like to have maximum confidence in your estimate. Unfortunately, it is impossible to provide a confidence interval in which the true mean is guaranteed to reside, short of making the interval infinitely wide. However, one generally wants the estimate to be precise, which means making the confidence internal as narrow as possible. This brings us to the cruel fact of the estimation business: the goal of making a useful (precise) estimate is in opposition to the goal of making a confidence interval with a large confidence level. One can always make a very precise estimate, but it will have a lower confidence level, or one may increase the confidence by making the interval larger.

So, given a data set, how do we construct a confidence interval? First, we need to choose the desired confidence level, for example, $\alpha = 0.95$, which is commonly used. The data set has a sample mean $\bar{X}$, sample standard deviation σ_X, and sample size n, which are at

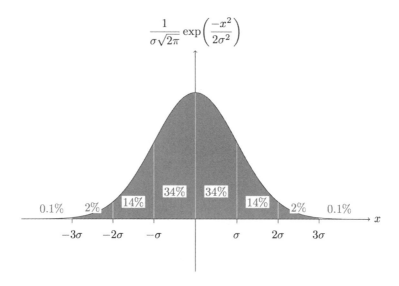

Figure 5.2: The normal probability density in terms of standard deviations from the mean.

our disposal. The Central Limit Theorem tells us that the distribution of sample means is normal (provided n is large enough) with mean μ and standard deviation $s = \sigma/\sqrt{n}$. One can calculate the deviation around the μ that defines a confidence interval using fancy integrals or computers. Figure 5.2 shows the probability of falling within a certain deviation from the mean for integer multiples of the standard deviation σ; for example, the probability of falling within one standard deviation of the mean for a normal random variable is approximately 68%. Luckily, these probability levels are the same for all normal distributions, so they can be used to calculate the confidence interval for any mean. For instance, the range of the normal random variable that contains 95% probability is approximately $(\mu - 1.96s, \mu + 1.96s)$, that is, between 1.96 times the standard deviation to the left of the mean to 1.96 the standard deviation to the right of the mean. So to construct the confidence interval, one needs the mean μ, the standard error s, and the 95% confidence level.

If you were not too bedazzled by different symbols, you might have noticed that there was a bait-and-switch in the argument

5.2. CENTRAL LIMIT THEOREM

above. The data set provides the sample mean $\bar{X}$, but the confidence interval requires the true mean μ. The Law of Large Numbers says that the former approaches the latter for large n, but they are not the same. However, there is no choice but to use the imperfect sample mean $\bar{X}$ as the central point for the confidence interval—because μ is what we are trying to estimate! Further, the standard error s actually depends on the true standard deviation σ, but we have to make do with the sample standard deviation σ_X, for the same reason: the true value is not available. (The topic of proper estimation of standard deviation is another kettle of fish, which I leave aside.) These approximations are not always sufficiently appreciated, but they can lead to notable discrepancies between the theoretical confidence level and the actual confidence level, which are illustrated in the computational assignments in section 5.5.

To summarize, here is the recipe for calculating a confidence interval for the true mean at a given confidence α based on a data set X of size n:

1. Compute the sample mean $\bar{X}$, which is the best estimate we have of the true mean μ.

2. Compute the sample standard deviation σ_X, which is the best estimate we have of the actual standard deviation σ.

3. Compute the standard error $s = \sigma_X/\sqrt{n}$.

4. Use a table or R to calculate the multiple called z_α for the desired confidence level α. For example, $z_{0.9} \approx 1.65$, $z_{0.95} \approx 1.96$, $z_{0.99} \approx 2.58$.

5. Build the confidence interval by adding and subtracting $z_\alpha s$ from the sample mean $\bar{X}$. For example, the 95% confidence interval is $\bar{X} \pm z_{0.95} s = \bar{X} \pm 1.96 \sigma_X/\sqrt{n}$.

Math exercises:

For the (small) data sets below, calculate the standard error and the confidence intervals at 90%, 95%, and 99% levels, using the z_α values provided in the recipe above.

Exercise 5.2.1. Data set of the population of the city of Chicago (in millions) in the past four census years (2010, 2000, 1990, 1980): 2.7 2.9 2.8 3.0.

Exercise 5.2.2. Data set of the numbers of the fish known as the blacknose dace (*Rhinichthys atratulus*) collected in six different streams in the Rock Creek watershed in Maryland: 76 102 12 55 93 98.

Exercise 5.2.3. Data set of tuberculosis incidence rates (per 100,000 people) in the five largest metropolitan areas in the United States in 2012: 5.2 6.6 3.2 5.5 4.5 (from the CDC: http://www.cdc.gov/tb/statistics/reports/2012/default.htm).

Exercise 5.2.4. Data set of ages of mothers at birth for five individuals: 19 20 22 32 39.

Exercise 5.2.5. Data set of ages of fathers at birth for five individuals: 22 23 25 36 40.

Exercise 5.2.6. Data set of the number of new mutations found on maternal chromosomes for five individuals: 9 10 11 26 15.

Exercise 5.2.7. Data set of the number of new mutations found on paternal chromosomes for five individuals: 39, 43, 51, 53, 91.

5.3 Confidence intervals for relative risk

Clinical trials to evaluate medical treatments or drugs usually proceed by dividing a group of people into two subgroups: one that receives the treatment and one that does not (the control group). If this is done in an unbiased random manner, it is called a randomized controlled trial (RCT), which is typically considered the best study design (for most purposes). The two groups are then compared for the outcomes, such as mortality or morbidity (illness). The comparison is usually done in the form of *relative risk*, or the ratio between the fractions of those with an undesirable outcome in one group and in the other. In an idealized case, if the relative risk is 1, there is no difference between the groups, and thus the

5.3. RELATIVE RISK

treatment has no effect. If the relative risk is not 1, then it makes a difference (either good or bad).

The astute reader has likely noted that this idealization has no practical value. Relative risk is almost never exactly 1, even if the treatment does nothing, due to chance alone. The actual relative risk from a study may be 1.12 or 0.96. Is this sufficiently different from 1 to say that the treatment has an effect? If only we had a way of computing a range of values to estimate the true value of a quantity ... But wait, we do! It's called a confidence interval.

The statistics necessary to compute confidence intervals for relative risk are different than what we have seen, because relative risk is not distributed normally. It turns out that its distribution is log-normal (under some assumptions). We will not delve into the details here, but the basic ideas are the same: choose the confidence level, and calculate the confidence interval based on the distribution (in this case log-normal) and the statistics of the data. The main question for a clinical trial is: Does this confidence interval include 1 or not? If not, then the treatment has an effect (whether positive or negative) at that confidence level.

Discussion questions:

The following discussion questions are based on the paper "Autism occurrence by MMR vaccine status among children with older siblings with and without autism" (Jain et al. 2015). Read the paper, think about it, and then discuss with your peers or colleagues.

Discussion 5.3.1. How are confidence intervals used in the study to conclude that there is no effect of MMR vaccination on development of autism spectrum disorders (ASD)?

Discussion 5.3.2. What are some limitations of the study? How does its design differ from a classic RCT?

Discussion 5.3.3. Does dividing children by whether they have a sibling with ASD make the study stronger?

5.4 Sampling and confidence intervals in R

5.4.1 simulated sampling

As we saw in section 4.4, R can be used to generate random numbers from a particular distribution, for example, the uniform distribution of real numbers between 0 and 1. This distribution differs from the discrete uniform distribution discussed in section 4.2, because the values of the random variable can be any real number between 0 and 1, so it is a continuous variable. The true mean of this distribution is 0.5, but what is the mean of a sample of numbers drawn from a uniform distribution? We can perform this numerical experiment using the R function runif(), which generates a specified number of random numbers from the uniform distribution. If you repeat the same command, you will get a different set of values, since they are generated randomly every time. The following script produces two random samples of size 10 and prints out the means of the samples:

```
sample <- runif(10)
print(sample)
```

```
##   [1] 0.2148808 0.2939575 0.8422431 0.8153124
##   [5] 0.5695632 0.3358997 0.6210980 0.3318065
##   [9] 0.1948192 0.3040391
```

```
mean(sample)
```

```
## [1] 0.452362
```

```
sample <- runif(10)
print(sample)
```

```
##   [1] 0.052141036 0.520944840 0.096708031
##   [4] 0.625973507 0.963039815 0.289936816
##   [7] 0.440226926 0.009591412 0.792221249
##  [10] 0.470470170
```

```
mean(sample)
```

```
## [1] 0.4261254
```

5.4. SAMPLING IN R

If you copy this script and run it yourself, you will obtain different numbers. If you run it several times, you will notice that the mean of sample of 10 values is prone to considerable volatility. This leads to two related questions: (1) How large does the sample need to be to obtain a good estimate of the true mean? (2) Given a random sample and its sample mean, what is the reasonable range of values for the estimate of the true mean? We addressed these questions theoretically in section 5.1. Here we investigate them by generating multiple random samples using R.

The script used to produce Figure 5.3 generates 100 random samples of size 10 (using the uniform random number generator), saves the means of the ten samples into a vector variable, and plots its histogram; then it does the same thing for 100 random samples of size 100. The histograms of sample means in Figure 5.3 demonstrate the effect of sample size that we previously discussed theoretically. As a smaller sample size, the sample means vary more widely; in other words, the distribution of sample means has a larger variance for smaller sample sizes. Although the sampling process is random, and every time the script is run it produces a new set of sample means, in the figure you can see clearly that sample means for samples of size 10 are much more spread out than those for sample size 100.

```
numsamples <- 100
samplemeans <- rep(NA, numsamples)
samplesize <- 10
for (i in 1:numsamples) {
    sample <- runif(samplesize)
    samplemeans[i] <- mean(sample)
}
```

The Central Limit Theorem predicts that for large sample size n, the distribution of sample means is close to the normal distribution. To demonstrate this, the scripts below generate 1000 samples of size 200 and 500 from the uniform random variable and plot the histograms of sample means shown in Figure 5.4. In addition to the histograms, we also overlay plots of the normal distribution with mean $\mu = 0.5$ and standard deviation $s = \sigma/\sqrt{n}$, where σ is the

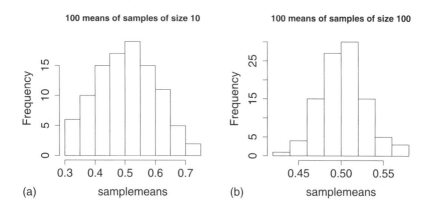

Figure 5.3: Distribution of means of samples drawn from the uniform distribution for different sample sizes: 10 (a) and 100 (b).

standard deviation of the continuous uniform distribution between 0 and 1, which happens to be $1/\sqrt{12}$. Notice that for sufficiently many samples (1000), the histograms are clearly bell-shaped, as predicted, despite the random samples being generated from the uniform distribution. The normal distribution curve matches the distribution of samples of size 1000 somewhat better than those of size 100, but the difference is not dramatic. The most important difference is in the spread of the sampling distribution, which is noticeably smaller for the larger sample size. This illustrates how much more accurate an estimate of the mean from sample of size 1000 is compared to sample of size 100.

5.4.2 computing confidence intervals

Now let us compute a confidence interval for the mean based on a sample, following the recipe given at the end of section 5.2. To do this, we will use the function qnorm(), which is the inverse normal density function. For a given probability level p, the function returns the value x of the standard normal random variable with mean $\mu = 0$ and standard deviation $\sigma = 1$, such that the total probability of the distribution being less than x is equal to p. For instance, since the standard normal is a symmetric distribution with mean 0, the

5.4. SAMPLING IN R

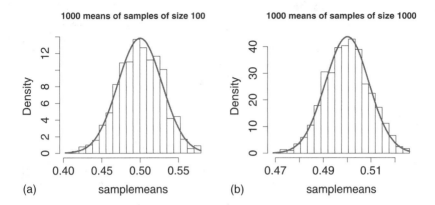

Figure 5.4: Central Limit Theorem in action: comparison of histograms of sample means and the normal distribution: (a) a histogram of means of samples of size 100, (b) one of means of samples of size 1000.

probability of the random variable being below 0 is 0.5, as shown by the command in the script below. The second command in the script calculates the value of the standard normal so that 95% of the probability is to its left.

```
qnorm(0.5)
# the value that divides the density function in two

## [1] 0

qnorm(0.95)
# the value such that 95% of density is to its left

## [1] 1.644854
```

The function qnorm() is necessary for computing the z_α values for a given confidence value α. The z_α is the value of the standard normal such that the random variable z has probability α of being within z_α of the mean $\mu = 0$; in other words, it requires that the probability of being outside of the interval $(-z_\alpha, z_\alpha)$ is $1 - \alpha$.

The function `qnorm(p)` returns the value x so that the standard normal has probability p of being less than x; in other words, that probability being greater than x is $1 - p$. There are two ways of being outside the interval $(-z_\alpha, z_\alpha)$: less than $-z_\alpha$ (called the left tail of the distribution) and greater than z_α (called the right tail of the distribution). The values z_α, by definition, represent tails of $(1 - \alpha)/2$, because there are two tails, and the normal distribution is symmetric.

The script below generates a sample from the uniform distribution, computes its mean, standard deviation, and the standard error. Then it calculates the value z_α based on the defined α value, and calculates the variables `right` and `left`, the respective boundaries of the confidence interval for the true mean. The script shows confidence intervals calculated from a sample of size 10 and a sample of size 100.

```
size <- 10   # sample size
alpha <- 0.95   # significance level
sample <- runif(size)
s <- sd(sample)/sqrt(size)   # standard error
z <- qnorm((1 - alpha)/2)   # z-value
left <- mean(sample) + s * z
right <- mean(sample) - s * z
print(right)

## [1] 0.7154246

print(left)

## [1] 0.3850585

size <- 100   # sample size
sample <- runif(size)
s <- sd(sample)/sqrt(size)   # standard error
z <- qnorm((1 - alpha)/2)   # z-value
right <- mean(sample) + s * z
left <- mean(sample) - s * z
print(right)
```

```
## [1] 0.4771531
```

```
print(left)
```

```
## [1] 0.5938232
```

Every time you run this script, it will produce new samples and therefore different confidence intervals, but most of the time (theoretically, 95% of the time, although that is not practically true) the confidence interval will contain the true mean of 0.5. As you would expect, the confidence interval for sample size 10 is much wider than the one for sample size 100. To obtain a reliable estimate of the true mean, you must obtain a sufficient number of measurements.

5.5 Computational projects

In this project you will explore random sampling using the R random number generators and observe how much the sample means differ from the true mean. You will generate uniform random numbers between 0 and 1 using the function `runif()`.

Tasks

1. Generate 100 samples of size 20 and save their mean values to a new vector variable. (a) Report the mean value of the 100 sample means; (b) report the standard deviation of the 100 sample means; (c) plot a histogram of the 100 sample means; d) what do you expect to be the mean of the distribution of sample means? How close is your simulated result to the expected value?

2. Generate 100 samples of size 180 and save their mean values to a new vector variable. Report (a) the mean value of the 100 sample means; (b) the standard deviation of the 100 sample means; (c) plot a histogram of the 100 sample means. (d) what do you expect to be the mean of the distribution of sample means? How close is your simulated result to the expected

value? (e) Compare the results of this simulation with the results for sample size 20.

3. What is the minimum sample size at which about 90% of random samples have means within 0.05 of the true mean of the uniform random variable between 0 and 1? Try different sample sizes, generate 1000 random samples, plot histograms of their sample means, and estimate how many are outside the desired range. Since this is a random experiment, you will get different results every time you run it, so I am only asking for an approximation. Report: (a) which sample sizes you tried; (b) approximately how many out of 1000 were outside the desired range; (c) your answer to the question.

Part II

Relationship between two variables

Chapter 6

Independence of random variables

> *Unconnected and free,*
> *No relationship to anything.*
> —They Might Be Giants, *Unrelated Thing*

In the first part of the book you learned how to describe data sets and probability distributions of random variables. So far we have not discussed how two or more variables may influence one another. The next four chapters are devoted to relationships between two variables. Many experiments in biology result in observations that naturally fall into a few categories, for example: sick or healthy patients, or presence or absence of a mutation. The resulting data sets are called *categorical*. Unlike *numerical* data sets that we will investigate in Chapters 8 and 9, they are not usually represented by numbers. Although it is possible, for instance, to denote mutants with the number 1 and wild type with 0, such designation does not add any value. Categorial variables require different tools for analysis than do numerical ones; one cannot compute a linear regression between two categorical variables, because there is no meaningful way to place categories on axes. In this chapter you will learn to do the following:

1. define conditional probability;
2. define independence for events and random variables;
3. use a categorical data table in a statistical test to decide whether two variables are independent;
4. enter a contingency table as a matrix in R; and
5. calculate the uncertainty of this decision, based on the data set using R.

6.1 Categorical data sets with two variables

What kind of relationship can there be between categorical variables? It cannot be expressed in algebraic form, because without numerical values we cannot talk about a variable increasing or decreasing. Instead, the question is, does one variable being in a particular category have an effect on which category the second variable falls into? Let us say you want to know whether the age of the mother has an effect on the child having trisomy 21 (i.e., Down syndrome), a genetic defect in which an embryo receives three chromosomes 21 instead of the normal two. The age of the mother is a numerical variable, but it can be classified into two categories: less than 35 and 35 or more years of age. The trisomy status of a fetus is clearly a binary, categorical variable: the fetus either has two chromosomes 21 or three.

The data are presented in a two-way or *contingency table*, which is a common way of presenting a data set with two categorical variables. The rows in such tables represent different categories of one variable, the columns represent the categories of the other, and the cells contain the data measurements of their overlaps. Table 6.1 shows a contingency table for the data set on Down syndrome and maternal age, in which the rows represent the two categories of maternal age and the columns represent the presence or absence of the syndrome. Each internal cell (as opposed to the total counts on the margins) corresponds to the number of measurement where both variables fall into the specified category; for instance, the number of fetuses with DS and a mother under 35 is 28.

6.2. MATHEMATICS OF INDEPENDENCE

Maternal age/ Down syndrome	No DS	DS	Total
Under 35 years	29,806	28	29,834
35 years or above	8,135	64	8,199
Total	37,941	92	38,033

Table 6.1: Contingency table for maternal age and incidence of Down syndrome. Numbers represent counts of patients belonging to both categories in the row and the column. DS = Down syndrome. From Malone et al. (2005).

Once the data are organized into a contingency table, we can address the main question stated above: does the age of the mother have an effect on whether a fetus inherits three chromosomes 21? Perhaps the first approach that suggests itself is to compare the fraction of mothers carrying a fetus with DS for the two age categories. In this case, the fraction for the under-35 category is $28/29834 \approx 0.00094$, while for the 35-and-over category the fraction is $64/8199 \approx 0.0078$. The two fractions are different by almost a factor of 10, which certainly suggests a real difference between the two categories. However, all data contain an element of randomness and a pinch of error; thus we need a quantifiable way of deciding what constitutes a real effect. This leads us to the questions posed and addressed in this chapter.

1. Based on the data, how strong is the effect of advanced maternal age on the prevalence of DS?

2. Is the observed effect real, or is it due to random variation sampling?

6.2 Mathematics of independence

6.2.1 conditional probability and information

Let us return to the abstract description of probability first introduced in section 4.1. There we used the notion of sample space and its subsets, called events, to describe collections of experimental outcomes. Suppose that you have some information about a random experiment that restricts the possible outcomes to a particular subset (event). In other words, you have ruled out some outcomes, so the only possible outcomes are those in the complementary set. This will affect the probability of other events in the sample space, because your information may have ruled out some of the outcomes in that event as well.

Definition 6.1. For two events A and B in a sample space Ω with a probability measure P, the *probability of A given B*, called the *conditional probability*, is defined as

$$P(A|B) = \frac{P(A \cap B)}{P(B)}$$

where $A \cap B$ is the intersection of events A and B, also known as "A and B" which is the event that consists of all outcomes that are in both A and B.

In words, given the knowledge that an event B occurs, the sample space is restricted to the subset B, which is why the denominator in the definition is $P(B)$. The numerator is all the outcomes we are interested in, which is A, but since we are now restricted to B, the numerator consists of all the elements of A that are also in B, or $A \cap B$. The definition makes sense in two extreme cases: if $A = B$ and if A and B are mutually exclusive:

- $P(B|B) = P(B \cap B)/P(B) = P(B)/P(B) = 1$ (probability of B given B is 1).

- If $P(A \cap B) = 0$, then $P(A|B) = 0/P(B) = 0$ (if A and B are mutually exclusive, then probability of A given B is 0).

There are some common misunderstandings about conditional probability, which are usually the result of discrepancies between everyday word usage and precise mathematical terminology. First, the probability of A given B is not the same as probability of A and B. These concepts seem interchangeable, because the statement "what are the odds of finding a tall person who likes tea?" is hard to distinguish from "what are the odds that a person who is tall likes tea?" The difference in these concepts can be illustrated using a Venn diagram, shown in Figure 6.1. Based on the probabilities

6.2. MATHEMATICS OF INDEPENDENCE

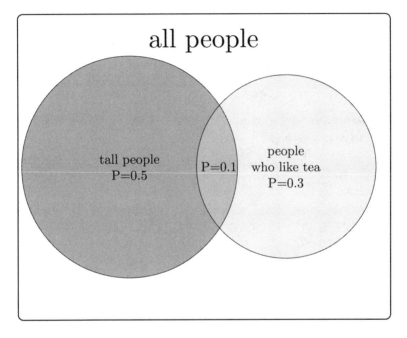

Figure 6.1: A Venn diagram of the sample space of all people with two events: tall people (A; dark gray) and those who like tea (B; light gray) with probabilities of A, B, and their intersection ($A \cap B$, medium gray area) indicated.

indicated there, the probability of randomly selecting a person who is both tall and likes tea is $P(A \cap B) = 0.1$, while the probability that a tea drinker is tall is $P(A|B) = 0.1/0.3 = 1/3$, which are different values.

A similar misconception is to be cavalier about the order of conditionality. In general, $P(A|B) \neq P(B|A)$, except in special cases. Going back to the illustration in Figure 6.1, the probability that a tea drinker is tall $P(A|B) = 1/3$ is the different than the probability that a tall person is a tea drinker $P(B|A) = 0.1/0.5 = 0.2$. One must take care when interpreting written statements to carefully distinguish what is known a priori and what remains under investigation. In the statement $P(A|B)$, B represents what is known, and A represents what is still to be investigated.

Example. Let us return to the data set discussed in section 6.1. Table 6.1 describes a sample space with four outcomes and several different events. One can calculate the probability of a fetus having Down syndrome (event) based on the entire data set of 38,033 mothers and 92 total cases of Down syndrome, so the probability is $92/38,033 \approx 0.0024$. Similarly, we can calculate the probability of a mother being older than 35 as $8,199/38,033 \approx 0.256$.

Now we can calculate the conditional probability of a mother older than 35 having a Down syndrome (DS) fetus, but first we have to be clear about what information is known and what is not. If the age of the mother is known to be over 35 (mature age, or MA), then we calculate $P(DS|MA) = 64/8,199 \approx 0.008$. Notice that the denominator is restricted by the information that the mother is over 35, and thus only women in that category need to be considered for the calculation.

In contrast, if we have the information that the fetus has DS, we can calculate the reversed conditional probability: that is, what is the probability that a fetus with DS has a mother above age 35? $P(MA|DS) = 64/92 \approx 0.7$. Notice that in both calculations the numerators are the same, since they both are the intersection between the two events, but the denominators are different, because they depend on which event is given.

6.2. MATHEMATICS OF INDEPENDENCE

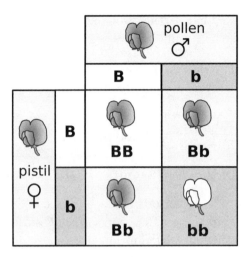

Figure 6.2: Punnet square of a cross of two heterozygous pea plants showing the possible genotypes and phenotypes of offspring. Figure by Madprime in public domain via Wikimedia Commons.

Math exercises:

Figure 6.2 shows a table of genotypes from the classic Mendelian experiment with genetics and color of pea flowers. The parents are both heterozygous, meaning each has a copy of the dominant (purple; shown as gray in the figure) allele B and the recessive (white) allele b. The possible genotypes of offspring are shown inside the square, and all four outcomes have equal probabilities. Based on this information, answer the following questions.

Exercise 6.2.1. What is the probability of an offspring having purple flowers? White flowers?

Exercise 6.2.2. What is the probability of an offspring having genotype BB? Genotype Bb? Genotype bb?

Exercise 6.2.3. What is the probability of an offspring having genotype BB, given that its flowers are purple?

Exercise 6.2.4. What is the probability of an offspring having genotype Bb, given that its flowers are purple?

Exercise 6.2.5. What is the probability of an offspring having genotype BB, given that its flowers are white?

Exercise 6.2.6. What is the probability of an offspring having genotype Bb, given that its flowers are white?

Exercise 6.2.7. What is the probability of an offspring having purple flowers, given that its genotype is BB?

6.2.2 independence of events

We first encountered the notion of independence in Chapter 4, where two events were said to be independent if they did not "affect each other." The mathematical definition uses the language of conditional probability to make this notion precise. It says that A and B are independent if given the knowledge of A, the probability of B remains the same, and vice versa.

Definition 6.2. Two events A and B are *independent* if $P(A|B) = P(A)$, or equivalently, if $P(B|A) = P(B)$.

Independence is not a straightforward concept. It may be confused with mutual exclusivity, as one might surmise that if A and B have no overlap, then they are independent. That however, is false by definition, since $P(A|B)$ is 0 for two mutually exclusive events. The confusion stems from thinking that if A and B are non-overlapping, then they do not influence each other. But the notion of influence in this definition is about information; so of course if A and B are mutually exclusive, the knowledge that one of them occurs has an influence of the probability of the other one occurring.

A useful way to think about independence is in terms of fractions of outcomes. The probability of A is the fraction of outcomes out of the entire sample space that is in A, while the probability of A given B is the fraction of outcomes in B that are also in A. The definition of independence equates the two fractions; therefore, if A occupies $1/2$ of sample space, for A and B to be independent, events in A must constitute $1/2$ of the event B. In Figure 6.1, the fraction of tall people is 0.5 of the sample space, but the fraction of tea drinkers

6.2. MATHEMATICS OF INDEPENDENCE

who are tall is $0.1/0.3 = 1/3$. Since the two fractions are different, A and B are not independent.

Math exercises:

Consider three examples of events and their intersections shown in Figure 6.3.

Exercise 6.2.8. Based on the two non-overlapping (mutually exclusive) events, calculate the conditional probability $P(A|B)$, and compare it with $P(A)$. Are A and B independent?

Exercise 6.2.9. Based on the two partially overlapping events, calculate the conditional probability $P(A|B)$, and compare it with $P(A)$. Are A and B independent?

Exercise 6.2.10. Based on the two completely overlapping events, calculate the conditional probability $P(A|B)$ and compare it with $P(A)$. Are A and B independent?

6.2.3 calculation of expected frequencies

The definition of independence is abstract, but it has a direct consequence of great computational value. From the definition of conditional probability, $P(A|B) = P(A \cap B)/P(B)$, and if A and B are independent, then $P(A|B)$ can be replaced with $P(A)$, leading to the expression $P(A) = P(A \cap B)/P(B)$. Multiplying both sides by $P(B)$ gives us the formula called the *product rule*, which states that two events are independent if and only if (Feller 1968):

$$P(A \cap B) = P(B)P(A) \tag{6.1}$$

The product rule is extremely useful for computing probability distributions of complicated random variables. Recall that the binomial distribution, which we saw in section 4.2, is based on a string of n Bernoulli trials that are independent of one another, which allows the calculation of the probability of a string of successes and failures, or heads and tails, and so forth. In practice, independence between processes is rarely true in the idealized mathematical sense.

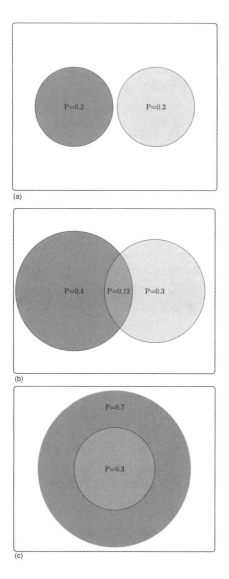

Figure 6.3: Three examples of two events inside a sample space with different probabilities: mutually exclusive (a), partially overlapping (b), and completely overlapping (c).

6.2. MATHEMATICS OF INDEPENDENCE

However, computing the probability of two random variables without independence is extremely difficult, so it is useful to make the independence assumption and then test it against the data. If it stands up, you have a good predictive model, and if it does not, you have learned that two processes are somehow linked, which is very useful information.

The product rule enables us to extend the notion of independence from events to variables. The concept of independence is the same in both contexts, since the probability of a value x of a random variable X corresponds to the probability of the event that gets mapped to x by the variable. To make independence applicable to variables, the condition must hold true for all possible values of both random variables. That way, knowing the value of one variable has no effect on the probability of the other. To make it simpler to calculate, we will use the product rule as the equivalent condition for independence:

Definition 6.3. Two random variables X and Y are *independent* if for all possible values of x and y

$$P(X = x \cap Y = y) = P(X = x)P(Y = y)$$

This allows us to address one of the questions posed at the beginning of the chapter: how can one measure whether a data set has independent variables? The definition allows us to calculate what we would expect if the variables were independent. Given a data set in the form of a contingency table, such as Table 6.1, we can first calculate the probabilities of the two variables separately, and then from that we predict the probabilities of the two variables together.

Example. Let us calculate the expected probabilities and frequencies of Down syndrome in pregnant women in the two age categories shown in Table 6.1. First, compute the probabilities of having Down syndrome (and not having it), based on all the pregnancies in the data set: $P(DS) = 92/38033 \approx 0.002419$; the complementary probability is $P(no\ DS) = 1 - P(DS)$. Similarly, we can calculate the probability that a pregnant woman is 35 or over, based on the entire data set (let's denote this event MA for "mature age").

$P(MA) = 8199/38033 \approx 0.21558$; the complementary probability is $P(YA) = 1 - P(MA)$ (YA stands for "young age").

These separate probabilities were calculated from the data, and now we can use them to calculate the predicted probabilities of different outcome, based on the assumption of independence. The probability of a mature-age woman having a pregnancy with Down syndrome, based on the product rule is $P(MA\&DS) \approx 0.0024 \times 0.216 = 0.000518$. Similarly, we can calculate the probabilities of the other three outcomes: $P(YA\&DS) \approx 0.0019$; $P(MA\&no\ DS) \approx 0.2156$; $P(YA\&no\ DS) \approx 0.782$.

These probabilities are the predictions based on the assumption that the two variables are independent. To compare the predictions with the data, we need to take one more step: convert the probabilities into counts, or frequencies of each occurrence. Since the probability is a fraction out of all outcomes, to generate the predicted frequency, we need to multiply the probability by the total number of data points (in this case the number of pregnant patients). The results of this calculation are seen in Table 6.2 with expected frequencies shown instead of experimental observations.

Maternal age/ DS	No DS	DS	Total
under 35 years	29,761.8	72.2	29,834
35 years or above	8,179.2	19.8	8,199
Total	37,941	92	38,033

Table 6.2: Expected frequencies of Down syndrome for two different age groups of mothers, assuming that maternal age and the occurrence of Down syndrome are independent. DS = Down syndrome.

Notice that expected frequencies do not need to be integers, because they are the result of prediction and not a data measurement. Now that we have a prediction, we can compare it with the measurements in Table 6.1. The numbers are substantially different, and we can see that the predicted frequency of Down syndrome for women under 35 is larger than the frequencies for women at or above 35, due to the larger fraction of patients in the younger age group. However, how certain are we that the difference is not due to chance, and is in fact is the result of a relationship between the

6.3. TESTING FOR INDEPENDENCE

two variables? To answer this question, we again leave the theoretical confines of probability and venture on an excursion into the wonderful and wild world of statistics.

6.3 Testing for independence

6.3.1 hypothesis testing

The scientific method is based on *testing hypotheses* and deciding whether to reject them. To do this, scientists formulate an idea (hypothesis), then accumulate data that can challenge it, and if the data contradict the hypothesis, they discard it (the hypothesis, not the data!). No hypothesis in science is ever proven in an absolute sense, which is why science is fundamentally different from mathematics. A hypothesis that has survived many tests and has been found to be consistent with all available observations becomes a *theory*, like the theory of gravity or of evolution. But unlike a theorem, a scientific theory is not certain, and if solid evidence were to surface that contradicts Netwon's gravitational theory, it would be falsified and thrown out (again, the theory, not the evidence).

The hypothesis to be tested is usually called the *null hypothesis*, which helpfully rhymes with "dull," because it represents the lack of anything interesting, essentially the default state of the system. To reject the null hypothesis, the data has to be substantially different from what is expected as default. For instance, medical tests have the null hypothesis that the patient is normal/healthy, and only if the results are substantially different from normal is the patient considered to be ill. Another common example is the criminal justice system: a defendant on trial undergoes a binary test where the null hypothesis is innocence. Only if the prosecutor's evidence is strong—that is, shows guilt beyond a reasonable doubt—is the null hypothesis rejected and the defendant found guilty.

Tests are binary, in that there are only two possible decisions: to reject the hypothesis or to not reject it. We can never truly accept a hypothesis as true, due to the impossibility of perfect knowledge of the world. The decision to reject a hypothesis is called a *positive* test result, which seems backward, but remember that the default or null hypothesis is a lack of anything unusual or interesting, so if the data are different from default, it is called a positive result. The decision to not reject the null hypothesis is called a *negative* test result. You are probably familiar with this in a medical context: if you've ever been tested for a disease, you know that a negative result is good news!

Hypothesis testing is also binary on the front end: if we had perfect knowledge, we could say whether the hypothesis is true or not. Ideally, we want the test to reject a false null hypothesis and to not reject a true null hypothesis. These results are called, respectively, a *true positive* and a *true negative*. In the language of probability, they are defined as follows.

Definition 6.4. For a hypothesis test, the probability of a *true positive* (TP) result is the probability of a positive result for a false hypothesis, and the probability of a *true negative* (TN) result is the probability of a negative result for a true hypothesis, or in mathematical notation:

$$P(TP) = P(Pos\&F); \ P(TN) = P(Neg\&T)$$

Conversely, the probability of a *false positive* (FP) result is the probability of a positive result for a true hypothesis, and the probability of a *false negative* (FN) result is the probability of a negative result for a false hypothesis, or in mathematical notation:

$$P(FP) = P(Pos\&T); \ P(FN) = P(Neg\&F)$$

Test result	Hypothesis false	Hypothesis true
Positive	TP	FP
Negative	FN	TN

Table 6.3: The four possible results of hypothesis testing. The values may be expressed either as counts or as fractions (probabilities).

These definitions are summarized in the Table 6.3. Based on these probabilities, one can compute the measures of quality of a given test, defined as follows.

6.3. TESTING FOR INDEPENDENCE

Definition 6.5. The *sensitivity* of a test is the probability of obtaining the positive result, given a false hypothesis; and the *specificity* of a test is the probability of obtaining the negative result, given a true hypothesis. There are two kinds of *testing error rates*: *type I error rate* is the probability of obtaining the positive result, given a true hypothesis (complementary to specificity); *type II error* rate is the probability of obtaining the negative result, given a false hypothesis (complementary to sensitivity). All four parameters (rates) of a binary test are summarized as follows:

$$Sen = \frac{TP}{TP+FN}; \quad Spec = \frac{TN}{TN+FP}$$

$$E1R = \frac{FP}{TN+FP}; \quad E2R = \frac{FN}{TP+FN}$$

Notice that knowledge of sensitivity and specificity determine the type I and type II error rates of a test, since they are complementary events. Of course, it is desirable for a test to be both very sensitive (reject false null hypotheses, detect disease, convict guilty defendants) and very specific (not reject true null hypotheses, correctly identify healthy patients, acquit innocent defendants), but no test is perfect, and sometimes it results in the wrong conclusion. This is where statistical inference comes into play: given some information about these parameters, a statistician can calculate the error rate in making different decisions.

Math exercises:

Table 6.4 shows the results of using X-ray imaging as a diagnostic test for tuberculosis in patients with known TB status. Use it to answer the questions below.

Exercise 6.3.1. Calculate the marginal probabilities of the individual random variables (i.e., the probability of positive and negative X-ray test results) and of TB being present and absent.

Exercise 6.3.2. Find the probability of a positive result given that TB is absent (false positive rate) and the probability of a negative result given that TB is absent (specificity).

Exercise 6.3.3. Find the probability of negative result given that TB is present (false negative rate) and the probability of a positive result given that TB is present (sensitivity).

Exercise 6.3.4. Find the probability that a person who tests positive actually has TB (probability of TB present given a positive result).

Exercise 6.3.5. Find the probability that a person who tests negative does not have TB (probability of no TB given a negative result).

Exercise 6.3.6. Assuming the test result and the TB status are independent, calculate the expected probability of both TB being present and a positive X-ray test.

Exercise 6.3.7. Under the same assumption, calculate the expected probability of both TB being absent and a positive X-ray test.

Test for TB	TB absent	TB present
Negative X-ray	1739	8
Positive X-ray	51	22

Table 6.4: Data for tuberculosis (TB) testing using X-ray imaging.

6.3.2 rejecting the null hypothesis

Hypothesis testing is one of the most important applications of statistics. A lot of people think of statistics as a collection of tests to be used for different hypotheses, which is too simplistic, but different tests do occupy a large fraction of statistics books. In this book we will only dip a toe into hypothesis testing, and will primarily approach it in a probabilistic (model-centered) way rather than from a statistical (data-centered) viewpoint. Probability allows us to calculate the sensitivity and specificity of a test for a given null hypothesis, provided the hypothesis is simple enough and the data are sampled correctly. Here is a simple example.

6.3. TESTING FOR INDEPENDENCE   141

Example: Testing whether a coin is fair. Suppose we want to know whether a coin is fair (has equal probabilities of heads and tails) based on a data set of several coin tosses. How much evidence do we need to reject the hypothesis of a fair coin with a small chance of making a type I error? What is the corresponding chance of making a type II error (i.e., not detecting an unfair coin)?

Let us first consider a data set of two coin tosses. If one is heads and one is tails, it's obvious we have no evidence to reject the null hypothesis. But what if both times the coin landed heads? The probability of this happening for a fair coin is 1/4, which means that if you reject the null hypothesis based on the evidence, your probability of committing a type I error is 1/4. However, it is difficult to answer the second question about making a type II error, because to do the calculation, we need to know something about the probability of heads or tails. The hypothesis being false only means that the probability is not 1/2, but it could be anything between 0 and 1.

Let us see how this test fares for a larger sample size. Suppose we toss a coin n times, and if all n come up heads, then we reject the hypothesis that the coin is fair. A fair coin will come up all heads with probability $1/2^n$, so that is the rate of false positives for this test. For example, if a coin came up heads ten times in a row, there is only a 1/1024 probability that this is the result of a fair coin, so the probability of making a type I error is less than 0.1%. Is this careful enough? This question cannot be answered mathematically—it depends on your sense of acceptable risk of making a mistake. Notice that if you decide to use a stringent criteria for rejecting a null hypothesis, you will necessarily end up not rejecting more false hypotheses. Such is the fate of us mortals, dealing with imperfect information in an uncertain world.

This leads us to an important new idea: the probability that a given data set is produced from the model of the null hypothesis is called the *p-value* of a test. In the example of coin tosses we just studied, the *p*-value was $p = 1/2^n$. However, what if the data had 9 heads out of 10 tosses? The *p*-value then would be the probability of obtaining 9 or 10 heads out of 10. This is because to compute the probability of making a false positive error, we consider all cases that could have produced the result that is as different from expectation, or even further from expectation (in this case, 5 heads out of 10) than the data (Whitlock and Schluter 2008).

Definition 6.6. For a given data set D and a null hypothesis H_0, the *p-value* is defined as the probability of obtaining a result as far or farther from expectation than the data, given the null hypothesis.

The p-value is the most used, misused, and even abused quantity is statistics, so please think carefully about its definition. One reason this notion is frequently misused is because it is tempting to conclude that the p-value is the probability of the null hypothesis being true, based on the data. That is not true! The definition has the opposite direction of conditionality—we assume that the null hypothesis is true, and based on that calculate the probability of obtaining the data. There is no way (according to classical "frequentist" statistics) of assigning a probability to the truth of a hypothesis, because it is not the result of an experiment.

The simplest way to describe the p-value is that it is the likelihood of the hypothesis, based on the data set. This means that the smaller the p-value, the less likely the hypothesis is, and one can be more certain about rejecting the hypothesis. Alternatively, the p-value represents the probability of making a type I error (i.e., or rejecting the correct null hypothesis). These two notions may seem to be in conflict, but they tell the same story: if the hypothesis is likely, the probability of making a type I error is high.

6.3.3 the chi-squared statistic

Now we are ready to address the question of testing the independence hypothesis based on the table of observations and the calculated table of expected counts. To measure the difference between what is expected for a data table with two independent variables and the actual observations, we need to summarize these differences by single number. One can devise several ways of doing this, but the accepted measure is called the *chi-squared statistic*. It is defined as follows.

Definition 6.7. The *chi-squared statistic* (χ^2) for the independence test is calculated on the basis of a two-way table with m rows and n

6.3. TESTING FOR INDEPENDENCE

columns as the sum of the differences between the observed counts and the computed expected counts:

$$\chi^2 = \sum_i \frac{(Observed(i) - Expected(i))^2}{Expected(i)}$$

The number of degrees of freedom (d.f.) of chi-squared is $d.f. = (m-1)(n-1)$.

This number describes how far away the data are from what is expected for an independent data set. Therefore, the larger the chi-squared statistic, the larger the differences are between observed and expected frequency, and thus the less likely is the null hypothesis of independence. However, simply obtaining the χ^2 is not enough to say whether the two variables are independent. We need to translate the chi-squared value into the language of probability: what is the probability of obtaining a data set with a particular χ^2 value, if those two variables were independent?

This question is answered using the *chi-squared probability distribution*, which describes the probability of the random variable χ^2. Like the normal distribution discussed in section 5.2 it is a continuous distribution, because χ^2 can take any (positive) real value. In addition, the χ^2 distribution has an even more complicated functional form than the normal distribution, so I do not present it here, because it is not enlightening. I will also not share the derivation of the mathematical form of the distribution, as it is far outside the goals of this text. In practice, nobody computes either the chi-squared statistic or its probability distribution function by hand; instead, computers handle these chores. The chi-squared distribution has one key parameter, called the number of degrees of freedom (d.f.), which was defined above. Depending on d.f. the distribution changes; specifically, for more degrees of freedom the distribution moves to the right; that is, the chi-squared values tend to be larger.

The chi-squared distribution is used to determine the probability of obtaining a chi-squared statistic as at least as large as observed, based on the null hypothesis of independence. Figure 6.4 shows a plot of the chi-squared distribution, as well as the total probability to the right of an observed χ^2 (shaded area under curve). This

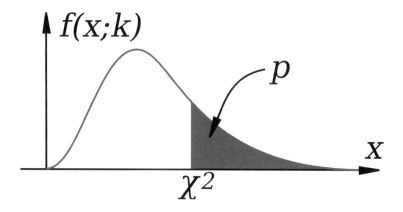

Figure 6.4: The chi-squared distribution $f(x;k)$ with k agreed of freedom is used to compute the p-value as the total probability p of obtaining a χ^2 value at least as far from 0 as observed. Image adapted from Inductiveload in public domain via Wikimedia Commons.

allows one to use it for the *chi-squared test* for independence between random variables by comparing the p-value obtained from the distribution (by a computer) against a number called the *significance* level, which is decided by humans. The significance value α is a threshold that the test has to clear to reject the null hypothesis: if the p-value is less than α, the independence hypothesis is rejected, otherwise it stands, although one can never say that the independence hypothesis is accepted.

No mathematical or statistical method exists for determining the appropriate significance level: it is entirely up to the users to decide how much risk of rejecting a true null hypothesis they are willing to tolerate. If you choose 0.01, that means you want the likelihood of the hypothesis to be less than 1% to reject it. This is entirely arbitrary, and using a rigid significance level to decide whether a hypothesis is true can lead to major problems, which we will discuss in Chapter 7.

Like all mathematical models, the chi-squared distribution relies on a set of assumptions. If the assumptions are violated, then the probability distribution does not apply, and the p-value does

6.4. HYPOTHESIS TESTING IN R

not reflect the actual likelihood of the hypothesis. Here are the assumptions:

- The data are from a simple random sample of the population, so that every subgroup of a certain size has an equal probability of being selected for the sample.

- The sample size must be sufficiently large; the exact number depends on the degrees of freedom.

- Expected cell counts cannot be too small; a rule of thumb is that they must be at least 5 for a 2-by-2 table.

- The observations are always assumed to be independent of one another; that is, one measurement has no effect on others. Note that this does not assume that the random variables are independent. (That's what we are testing!)

6.4 Hypothesis testing in R

R has many functions for different tests, including the chi-squared test. To use it, one first has to input a data set in the form of a two-way table, where each row represents the values of one random variable, and each column represents the values of the second random variable. The following script shows how to manually input a 2-by-2 contingency table into a matrix. In the matrix function, `ncol` stands for number of columns, and `nrow` for number of rows. Notice the order in which the numbers are put into the matrix: down the first column, then the second, and so forth. Type `help(matrix)` for more details. To access a specific element of the matrix, just like in vectors, R uses square brackets and two indices, first one for row, and second for column. Below are examples of R script accessing two elements of the matrix data defined above, and how to reference a particular element of the matrix.

```
data <- matrix(c(442, 514, 38, 6), ncol = 2, nrow = 2)
print(data)
```

```
##      [,1] [,2]
## [1,]  442   38
## [2,]  514    6

print(data[1, 2])

## [1] 38

print(data[2, 1])

## [1] 514
```

Based on a given data set, how likely is the hypothesis that the two random variables are independent? It is hard to determine this by hand (in the old days, you looked it up in a table of chi-squared values), but R will do it all for us: (1) calculate the expected counts, (2) compute the chi-squared value for the table, and (3) use the number of degrees of freedom and the chi-squared value to calculate the p-value of the independence hypothesis based on it. Use the chisq.test() function, and you will see output like this:

```
test.output <- chisq.test(data)
print(test.output)

## 
##  Pearson's Chi-squared test with Yates'
##  continuity correction
## 
## data:  data
## X-squared = 25.555, df = 1, p-value =
## 4.3e-07
```

The results are the chi-squared values, the number of degrees of freedom (which depends on the number of rows and columns in the two-way table), and the p-value. The p-value is used to decide whether to reject the hypothesis, because it represents the likelihood of the hypothesis, given the data. In this case, the p-value

6.5. INDEPENDENCE IN DATA SETS

is pretty small, so it seems relatively safe to reject the hypothesis of independence. To see the results of the hypothesis test, type print(test.output), and to access the p-value individually, use test.output$p.value.

Finally, we need to specify the significance level α for the hypothesis test. This refers to the probability of rejecting a true null hypothesis by random chance. For instance, if you reject the hypothesis at $\alpha = 0.05$ significance, you're accepting a 5% chance that you falsely rejected a correct hypothesis (type I error). Note that it says nothing about failing to reject an incorrect hypothesis (type II error).

6.5 Independence in data sets

6.5.1 maternal age and Down syndrome

Let us return to the data presented in section 6.1. We noted that the fractions of women in different age categories carrying fetuses with DS are different, but how certain are we that this is not a fluke? To test the hypothesis of independence, we input the data into R and then run the chi-squared test:

```
data <- matrix(c(29806, 8135, 28, 64), ncol = 2,
nrow = 2)
test.output <- chisq.test(data)
print(test.output)

##
##  Pearson's Chi-squared test with Yates'
##  continuity correction
##
## data:  data
## X-squared = 122.86, df = 1, p-value <
## 2.2e-16
```

Based on the calculations, we can answer the questions posed at the end of section 6.1:

1. Independence between the two variables means that the fraction of women with DS fetuses should be the same for all age groups (or vice versa).

2. To measure the extent of dependence of the two variables, calculate the chi-squared parameter, which is about 122.

3. This chi-squared parameter allows us to calculate the p-value and test the hypothesis. Here the p-value is tiny (the number is actually caused by machine error). Therefore, the hypothesis can be rejected with a very small risk of making an error.

6.5.2 stop-and-frisk and race

The practice of New York Police Department (NYPD) dubbed "stop-and-frisk" gives police officers to power to stop, question, and search people on the street without a warrant. Since the practice commenced in the early 2000s, it has generated controversy for several reasons. First, the fourth amendment to the U.S. Constitution limits the power of the state to detain and search citizens by mandating that officials first obtain a warrant based on "probable cause," while based on the Supreme Court interpretation, police are allowed to stop someone without a warrant, provided "the officer has a reasonable suspicion supported by articulable facts" that the person may be engaged in criminal activity. Exactly what these conditions mean and whether officers in NYPD always had reasonable suspicions before stopping is a legal matter, rather than a statistical one, and you can read what Federal Judge Scheindlin ruled in a trial on this matter (*New York Times*, 2014).

The second issue raised by stop-and-frisk is whether it violates the principle of equal protection under the law enshrined in the fourteenth amendment to the Constitution. The idea that the law and its agents should treat people of different backgrounds the same—that people can be punished for their actions, but not for who they are—is deeply rooted in American law and culture. Critics of stop-and-frisk charge that officers disproportionately stop and search people of African-American and Hispanic background and therefore violate their constitutional rights to equal protection. As part

6.5. INDEPENDENCE IN DATA SETS

of the Federal court trial, statistical evidence was introduced about the number of stops of New Yorkers of different racial backgrounds, how many of those stops resulted in the use of force, and how many uncovered evidence of criminal activity leading to an arrest. Let us analyze the data using our tools to address whether race and somebody being "stopped-and-frisked" are related.

The data in the summary of judge Scheindlin's decision is as follows: between 2004 to 2012, out of 4.4 million stops, 52% of the people stopped were black, 31% of the people stopped were Hispanic, and 10% of the people were white. The population of New York according to the 2010 census is approximately 23% black, 29% Hispanic, and 33% white. You may notice that the fractions are suggestive of a higher probability of stops for black people and lower probability of stops for white individuals, but we cannot use fractions to perform a chi-squared test, because actual counts are necessary to quantify the uncertainty in the testing.

Below I present data for counts only for the calendar year 2011 (Center for Constitutional Rights 2014), in the form of a contingency table with two variables: race/ethnicity and being stopped by police without a warrant. I have used the census population of New York (http://factfinder2.census.gov) and its breakdown by race (white only, black only, Hispanic, other). The data are presented in Table 6.5. They have been input in R and run through a chi-squared independence test.

	White	Black	Hispanic	Other	Total
Stopped	61,805	350,743	223,740	49,436	685,724
Not stopped	2,665,172	1,527,029	2,119,718	1,201,578	7,513,497
Total	2,726,977	1,877,772	2,343,458	1,251,014	8,199,221

Table 6.5: Number of New York residents subjected to stop-and-frisk and their racial identification.

```
data <- matrix(c(61805, 2665172, 350743, 1527029,
    223740, 2119718, 49436, 1201578), ncol = 4, nrow = 2)
test.output <- chisq.test(data)
print(test.output)
```

```
## 
##  Pearson's Chi-squared test
## 
## data:  data
## X-squared = 429040, df = 3, p-value <
## 2.2e-16
```

The results confirm what comparing the percentages suggested: the race of a person in NYC is not independent of whether they get stopped and frisked, with only a tiny probability that this disparity could have happened by chance. However, this is only the beginning of the analysis that experts performed for the court trial. Drawing conclusions about motives from the data is tricky, since two variables may be related without a causal connection. Defenders of the practice have argued that the racial disparities reflect differences in criminal activity. The data, however, show that only 6% of the stops result in arrests, and 6% more in court summons, so nearly all of those stopped and frisked were not engaged in criminal activity.

6.6 Computational projects

In this section you will use two data sets: one that you collect from your class, and one from marine biology research. In both cases, you will input the data into a matrix in R and then use the chi-squared function to test for independence between two variables.

6.6.1 thumb-on-top preference and sex

Collect data in you group of students on the relationship between two random variables: sex (male or female) and thumb-on-top preference (left or right). Ask everyone to determine whether they prefer to have the right or left thumb on top when they clasp their hands with fingers interlaced (the preference is usually very strong, if you try to clasp your hands the other way, it feels uncomfortable). Those who are male and prefer to have their right thumb on top raise their hands to be counted, then female students who prefer their right thumb on top, and so on, until you have filled the 2-by-2

6.6. COMPUTATIONAL PROJECTS

data table. (Students who don't identify as one of the traditional genders can pick either one.)

Tasks

1. Use the data set to construct a two-way table with handedness and thumb-on-top preferences as the two variables and assign it to an R matrix.

2. Use the R chi-squared function to calculate the p-value for the data set and decide whether to reject the hypothesis that sex and thumb on top preference are independent at significance levels $\alpha = 0.05, 0.01, 0.001$.

6.6.2 relationship between species and habitat

Habitat	G. moringa	G. vicinus
grass	127	116
sand	99	67
border	264	161

Table 6.6: Number of individuals of each species of eel sighted in each habitat.

The 2-way Table 6.6 contains data for sightings of two moray eels species of genus *Gymnothorax*, species *moringa* and *vicinus*, in different habitats in a reef in Belize. You will test the hypothesis that the species of eel and the habitat are independent.

Tasks

1. Put the data set in the two-way Table 6.6 into an R matrix.

2. Use the R chi-squared function to calculate the p-value for the data set and decide whether to reject the hypothesis that species and habitat are independent at significance levels $\alpha = 0.05, 0.01, 0.001$.

6.6.3 independence testing of simulated data

In this section you will generate random numbers to produce a simulated 2-by-2 contingency table. Suppose that you have two groups of people: one with genotype A and the other with genotype B. The question is: does this genotype predispose people to a disease? In other words, are the variables of genotype and disease linked?

You will generate a data set in which the true answer is known, so you can investigate how frequently the test is right and wrong. To do this, use the binomial random generator we first used in section 4.4. The code below generates two vectors of length `sample.size` containing 0s and 1s, in which 0 stands for normal and 1 indicates disease. The vector `dis.genA` contains the disease status of all people with genotype A, and `dis.genB` similarly contains the status for people with genotype B. The number of 0s and 1s in the two vectors are then tabulated using the `table()` function and are assigned to a data table.

```
sample.size <- 100
probA <- 0.2    # probability of disease for genotype A
probB <- 0.3    # probability of disease for genotype B
dis.genA <- rbinom(sample.size, 1, probA)
dis.genB <- rbinom(sample.size, 1, probB)
data.mat <- matrix(c(table(dis.genA), table(dis.genB)),
    nrow = 2, ncol = 2)
```

You are now ready to generate 2-by-2 tables for different scenarios and see how well the chi-squared test works for different probabilities of disease. Before doing the following tasks, think carefully: What are the variables in this test for independence? What does it mean for them to be independent?

Tasks

1. Generate data sets for genotype A and genotype B of 50 patients each with 0.5 probability of disease for both data sets. Plot the histograms of both data sets, and report whether they look similar. Place the counts into a data matrix, and run a chi-squared test on it. Does the test lead to rejection

6.6. COMPUTATIONAL PROJECTS

for the independence hypothesis at the 0.1 significance level? How about at 0.05? Based on how you generated the data sets, is the hypothesis actually true? Did the chi-squared test get the result right?

2. Repeat the code in the first task 100 times by adding a for loop around it. In each loop iteration, you generate the two datasets as before (both with probability of disease of 0.5), place the counts in a data matrix, and perform a chi-squared test for independence of genotype and disease. Report how many of the 100 chi-squared tests result in rejection of the null hypothesis at the 0.1 and 0.05 significance level by using test.output$p.value to put the p-values in a vector (e.g., p.vec), and report how many of the p-values are less than the significance level a (the command sum(p.vec<a) will do this for you). Since you know exactly whether the hypothesis is true, how many of the test conclusions are wrong for each significance level? Explain whether this agrees with the notion of p-value.

3. Now change the two data sets so they are different, and thus the data are not independent. For the first data set (genotype A), let the probability of disease be 0.4 and for the second data set (genotype B) let disease occur with probability 0.6. Count how many of the 100 chi-squared tests result in rejecting the null hypothesis at the 0.1 and 0.05 significance levels. Since you know exactly whether the hypothesis is true, how many of the test conclusions are wrong for each significance level? Explain whether this agrees with the notion of p-value.

4. Finally, let us generate data from very different probabilities of disease. For the first data set (genotype A), let the probability of disease be 0.2 and for the second data set (genotype B), let disease occur with probability 0.8. Count how many of the 100 chi-squared tests result in rejecting the null hypothesis at the 0.1 and 0.05 significance levels. Since you know exactly whether the hypothesis is true, how many of the test conclusions are wrong for each significance level? Explain whether this agrees with the notion of p-value.

Chapter 7

Bayes' amazing formula

> *Man can believe the impossible, but man can never believe the improbable.*
> —Oscar Wilde, *The Decay of Lying*

In Chapter 6 we defined the notion of independence and learned how to perform a statistical test to determine how likely a data set of two categorical variables is to have come from two independent random variables. This approach comes from the toolbox of classical "frequentist" statistics, which is taught in every statistics textbook in the world. It reduces statistical inference to a binary choice: reject or not reject the null hypothesis, based on the magic number called the p-value. However, this approach has deep problems, especially when applied mechanically and without understanding its limitations. Perhaps the most important limitation is that p-value based hypothesis testing does not incorporate any knowledge into its decision-making, aside from the given data. This may be reasonable at an early, exploratory stage of an experiment, but usually one has some *prior knowledge* about the likelihood of the hypothesis being tested. This knowledge cannot influence the data and the calculation of the p-value, of course, but it can have a dramatic effect on the interpretation, or *inference* one draws from the test. In this chapter you will learn to do the following:

1. explain the effect or prior knowledge on interpretation of an experimental result;

2. calculate the post-test probability based on the prior probability, test result, and the sensitivity and specificity of the test;

3. use conditional statements in R to simulate random decisions with a given probability; and

4. explain why conclusions based on binary p-value testing are frequently wrong.

7.1 Prior knowledge

Suppose that a patient walks into a doctor's office, the doctor orders a pregnancy test, and the results indicate that the patient is pregnant. The doctor consults the published sensitivity and specificity values (which were described in Chapter 6) to discover, for instance, that 99% of positive pregnancy tests are correct. The doctor goes back to congratulate the patient with impending motherhood. Sound very reasonable, doesn't it? Would it still sound reasonable if the doctor knew that the patient is a biological male?

This is a slightly absurd example, of course, but it neatly illustrates the central point of this chapter: prior knowledge has an effect on the inference from a test, no matter how small or large the p-value and the power of the test. If a patient is male, his *prior probability* of being pregnant is 0, and that is not changed by a test, no matter how accurate (of course, no test is 100% accurate). In this case, the positive pregnancy test must have been a false positive, even though it's unlikely, since the other possibility—that of a true positive—is impossible.

We all consider prior knowledge before coming to a conclusion. For instance, the credibility of a statement from a person very much depends on past performance: if the person is a habitual liar, you probably wouldn't put much stock in his or her words. In contrast, if a person is known to be trustworthy, you might take their statement seriously even if it is surprising. If your significant other has always been transparent and honest, even if you discover a suspiciously sexy message from someone else on his/her phone, you will listen to their explanation and consider alternative explanations, in other words, that this was a false positive. If they had abused your trust in the past, it's much more likely that this sexy text is actual evidence of cheating, and it's time to cut them loose!

In the context of science, the accumulation of knowledge is the basis of building scientific theories. Nothing in science is ever proven, unlike in mathematics; instead, different statements have different degrees of certainty based on past experience. A theory that has been tested for years—for instance, Newton's theory of gravity, with all experiments agreeing with its predictions (let's forget about relativistic corrections for a moment)—is considered to have a very high likelihood and is even called a law of nature. If experimental data came along that challenged Newton's law of gravity (as Einstein's relativity did), scientists would rightfully treat it much more skeptically than an experiment that agrees with prior experience. Carl Sagan summarized the effect of prior knowledge of evaluating evidence with the pithy phrase "extraordinary claims require extraordinary evidence"—that is, a claim that is highly unlikely based on past knowledge must be backed up by very strong data (extremely low p-value and high power).

7.2 Bayes' formula

In this section we formalize the process of incorporation of prior knowledge into probabilistic inference by going back to the notion of conditional probability discussed in section 6.2. First, if you multiply both sides of the expression definition in 6.1 by $P(B)$, then we obtain the probability of the intersection of events A and B in two different forms:

$$P(A \cap B) = P(A|B)P(B); \quad P(A \cap B) = P(B|A)P(A)$$

Second, we can partition a sample space into two complementary sets, A and $-A$, and then the set of B can be partitioned into two parts that intersect with A and $-A$, respectively, so that the probability of B is

$$P(B) = P(A \cap B) + P(-A \cap B)$$

Combine the expression for P $(A \cap B)$ from the calculation above with the expression for P(B) to obtain an important result called the *law of total probability* (Watkins n.d.):

$$P(B) = P(B|A)P(A) + P(B|-A)P(-A) \qquad (7.1)$$

It may not be clear at first glance why this is useful: after all, we replaced something simple ($P(B)$) with something much more complex on the right-hand side. You will see how this formula enables us to calculate quantities that are not otherwise accessible.

Example: Probability of a negative test result. Suppose we know that the probability of a patient having a disease is 1% (called the *prevalence* of the disease in a population), and the sensitivity and specificity of the test are both 80%. What is the probability of obtaining a negative test result for a randomly selected patient? Let us call $P(H) = 0.99$ the probability of a healthy patient and $P(D) = 0.01$ the probability of a diseased patient. Then

$$P(Neg) = P(Neg|H)P(H) + P(Neg|D)P(D) =$$
$$= 0.8 \times 0.99 + 0.2 \times 0.01 = 0.794$$

There is still more gold in the hills of conditional probability! Take the first formula in this section, which expresses the probability $P(A \cap B)$ in two different ways. Since the expressions are equal, we can combine them into one equation, and by dividing both sides by $P(B)$, we obtain what is known as *Bayes' formula*:

$$P(A|B) = \frac{P(B|A)P(A)}{P(B)}$$

The more useful version of Bayes' formula rewrites the denominator using the law of total probability (equation 7.1) to obtain

$$P(A|B) = \frac{P(B|A)P(A)}{P(B|A)P(A) + P(B|-A)P(-A)} \qquad (7.2)$$

7.2. BAYES' FORMULA

Bayes' formula gives the probability of A given B from probabilities of B given A and given $-A$, and the prior (baseline) probability of $P(A)$. This is enormously useful when it is easy to calculate the conditionals one way and not the other. Among its many applications is computing the effect of a test result with given sensitivity and specificity (conditional probabilities) on the probability of the hypothesis being true.

7.2.1 positive and negative predictive values

In reality, a doctor doesn't have the true information about the patient's health, but rather the information from the test and hopefully some information about the population where she is working. Let us assume we know the rate of false positives $P(Pos|H)$ and the rate of false negatives $P(Neg|D)$, as well as the prevalence of the disease in the whole population $P(D)$. Then we can use Bayes' formula to answer the practical question: if the test result is positive, what is the probability the patient is actually sick? This is called the *positive predictive value* of a test. The deep Bayesian fact is that one cannot make inferences about the health of the patient after the test without some prior knowledge, specifically, the prevalence of the disease in the population:

$$P(D|Pos) = \frac{P(Pos|D)P(D)}{P(Pos|D)P(D) + P(Pos|H)P(H)}$$

Example. Suppose the test has a 0.01 probability of both false positive and false negatives, and the overall prevalence of the disease in the population is 0.02. You may be surprised that from an epidemiological perspective, a positive result is far from definitive:

$$P(D|Pos) = \frac{0.99 \times 0.02}{0.99 \times 0.02 + 0.01 \times 0.98} = 0.67$$

This is because the disease is so rare, that even though the test is quite accurate, there are going to be a lot of false positives (about 1/3 of the time), since 98% of the patients are healthy.

We can also calculate the probability of a patient who tests negative of actually being healthy, which is called the *negative predictive value*. In this example, it is far more definitive:

$$P(H|Neg) = \frac{P(Neg|H)P(H)}{P(Neg|H)P(H) + P(Neg|D)P(D)} = \frac{0.99 \times 0.98}{0.99 \times 0.98 + 0.01 \times 0.02} = 0.9998$$

This is again because this disease is quite rare in this population, so a negative test result is almost guaranteed to be correct. In another population, where disease is more prevalent, this may not be the case.

Figure 7.1 illustrates all possibilities of a binary medical test: positive (P) or negative (N) for a patient who is either healthy (H) or diseased (D). The four outcomes correspond to the four outcomes of tests we saw in section 6.5: true positives are D&P, false positives are H&P, true negatives are H&N, and false negatives are D&N. This allows us to calculate the positive predictive value (PPV), which is the probability that a positive result is correct. For the patient on the left with disease prevalence of 0.1, $PPV = TP/(TP+FP) = 0.098/(0.098+0.045) \approx 0.685$. For the patient on the right with disease prevalence of 0.01, $PPV = TP/(TP+FP) = 0.0098/(0.0098+0.0495) \approx 0.165$. The exact same test has a higher PPV for a patient who has a higher prior probability of having a disease.

The negative predictive value (NPV) can be calculated in a similar manner: $NPV = TN/(TN+FN) = 0.855/(0.855+0.002) \approx 0.998$. For the patient on the right with disease prevalence of 0.01, $NPV = TN/(TN+FN) = 0.9405/(0.9405+0.0002) \approx 0.9998$. The exact same test has a higher NPV for a patient who has a lower prior probability of having a disease.

Math exercises:

In the problem set in section 6.3 you used the TB and X-ray test data in Table 6.4 to calculate the sensitivity and specificity of the test. Suppose that you know that the prevalence of TB in a population is $P(D) = 0.001$. Use the previously calculated sensitivity and

7.2. BAYES' FORMULA

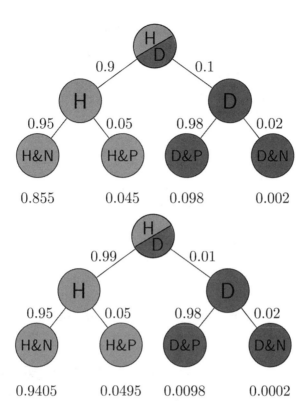

Figure 7.1: Probabilities of the four possible outcomes for patients with different disease prevalence using the same medical test with sensitivity (rate of true positives) of 0.98 and specificity (rate of true negatives) of 0.95: on the left is disease prevalence of 10%, and on the right prevalence of 1%. D = diseased; H = healthy; N = negative; P = positive.

specificity to answer the following questions about testing a patient from the population (not those used in the study in Table 6.4.)

Exercise 7.2.1. Using the law of total probability, calculate $P(Pos)$, the probability that a randomly chosen person from the population tests positive for the disease and $P(Neg)$, the probability that a randomly chosen person tests negative for the disease.

Exercise 7.2.2. Using Bayes' formula (equation 7.2), find the probability that a patient who tested positive has the disease

$P(D|Pos)$ and the probability that a patient who tested negative is healthy $P(H|Neg)$.

Exercise 7.2.3. Suppose the disease prevalence is $P(D) = 0.5$. Repeat the calculations to find the new $P(D|Pos)$ and $P(H|Neg)$.

7.3 Applications of Bayesian thinking

The essence of the Bayesian approach to statistics is that everything comes with a prior probability, or "odds," as Bayesians like to express it. There is usually some prior knowledge one has to assess the odds that a hypothesis is true before doing an experiment, which is called the *prior*. If you don't explicitly assume the prior, then you've assumed it implicitly. For example, the naive answer for the positive predictive value that ignores the prior prevalence for a test with 99% specificity and sensitivity is that it is also 99%. This assumes that the patient has equal prior probability of disease and health, as you can verify using Bayes' formula. Even in the absence of any data for a particular population, this is an unlikely assumption for most diseases. So the Bayesian advice is: assign odds to everything to the best of your knowledge so you don't get played for a sucker. For an excellent summary of the misuses of p-value and the Bayesian approach to interpretation of medical data, see (Goodman 1999).

7.3.1 when too much testing is bad

For decades, doctors recommended early cancer screening in a major public health effort to help reduce the mortality rate from cancer. This makes sense, because the prognosis is generally much better for cancer when it is detected early, since the tumor is small and has not yet metastasized. This approach has taken hold in the public imagination in the United States and has given us pink-ribbon campaigns aimed at breast cancer awareness and celebrities advising everyone to get tested early and often.

More recently, large-scale studies have shown that that preventive screenings do not necessarily improve the survival rates of cancer patients. One study from Canada (Miller et al. 2014) assigned almost 90,000 women randomly into two equal-sized groups by a

7.3. APPLICATIONS OF BAYESIAN THINKING

randomization process illustrated in Figure 7.2. One of the groups received yearly mammography screenings for 5 years, and one did not. The women were between the ages of 40 and 59. The study then tracked the participants for 25 years to see whether there was a difference in cancer mortality rates between the two groups.

The results may be surprising: the mortality rates were very similar, and in fact, slightly more women died of breast cancer in the mammography group than in the control group (180 vs 171). At the same time, more cases of invasive breast cancers were diagnosed in the mammography group, which may mean either that early treatment did not make a difference or that many of those cases were false positives.

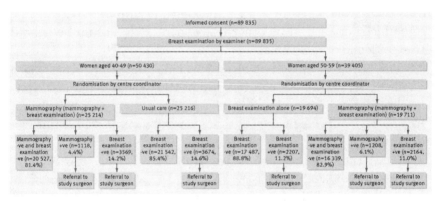

Figure 7.2: The study design for the large randomized screening trial of effectiveness of mammography screening. Figure from Miller et al. (2014) used by permission.

This may seem counterintuitive: isn't more information, however imperfect, better? Viewed from a Bayesian perspective, the results should be less than surprising. Since the prior probability of developing breast cancer in any given year is small, the positive predictive value of the test is likely low, and most of the positive results end up being false positives. A false positive result for breast cancer has a major negative impact on a person's life: it means more invasive testing, a lot of worrying, and sometimes unnecessary treatment with serious side effects. This does not mean, of course, that cancer screening is never useful, and I am in no way trying to offer medical advice. For patient populations with a higher prior probability, such screening tests may in fact provide a substantial benefit. But this once again underscores the importance of taking into account prior knowledge.

7.3.2 reliability of scientific studies

In 2005 John Ioannidis published a paper titled "Why Most Published Research Findings Are False" (Ioannidis , 2005b). The paper, as you can see by its title, was intended to be provocative, but it is based solidly on the classic formula of Bayes. The motivation for the paper came from the observation that too often in modern science, big, splashy studies that have been published can't be reproduced or verified by other researchers. What could be behind this epidemic of questionable scientific work?

The problem, as described by Ioannidis and many others, in a nutshell, is that the thoughtless use of traditional hypothesis testing leads to a high probability of false positive results being published. The Ioannidis paper outlines several ways in which this can occur.

First, there is the problem of prior knowledge. Too often, a hypothesis is tested, and if the resultant p-value is less than some arbitrary threshold (very often 0.05, an absurdly high number), then the results are published. However, if one is testing a hypothesis with low prior probability, a positive hypothesis test result is very likely a false positive. Very often, modern biomedical research involves digging through a large amount of information, like an entire human genome, in search for associations between different genes and a phenotype, like a disease. It is a priori unlikely that any specific gene is linked to a given phenotype, because most genes have very specific functions, and are expressed quite selectively, only at specific times or in specific types of cells. However, publishing such studies results in splashy headlines ("Scientists find a gene linked to autism!"), and so a lot of false positive results are reported, only to be refuted later, in much less publicized studies.

The second problem compounds the first one: multiple research groups studying the same phenomenon. This should be a good

7.3. APPLICATIONS OF BAYESIAN THINKING 165

thing, but it can lead to a higher volume of false positive results. Suppose that 20 groups are all testing the same hypothesis and are using the same p-value cutoff of 0.05 to decide whether their results are "significant." Even if the null hypothesis is true (i.e., there is no effect), 1 out 20 groups is likely to obtain a p-value less than 0.05, simply by random variation. What do you think that group will do? Yes, they should compare its results with the other groups, or try to repeat the experiment multiple times. But repeating experiments is costly and boring, and telling your competitors about your results can lead to your getting scooped. Better publish fast!

The third problem is even more insidious: bias in the experimental work, either conscious or not. Some of it may be due to experimental design, like biased sampling, or defective instrumentation—no experiment is perfect. One big violation of good experimental design is known as p-value fishing: repeating the experiment, or increasing the sample size, until the p-value is below the desired threshold, and then stopping the experiment. Using such defective design dramatically lowers the likelihood that the result is a true positive. And of course there is actual fraud, or fudging of data, which contributes to some bogus results.

Ioannidis performed basic calculations of the probability that a published study is true (i.e., that a positive reported result is a true positive), and how it is affected by pre-study (prior) probability, the number of studies conducted on the same hypothesis, and the level of bias. His prediction is that for a fairly typical scenario (e.g., pre-study probability of 10%, ten groups working simultaneously, and a reasonable amount of bias), the probability that a published result is correct is less than 50%. This effect is shown in Figure 7.3, taken from his paper. He then followed up with another paper (Ioannidis 2005a) that investigated 49 top-cited medical research publications over a decade and looked at whether follow-up studies could replicate the results. He found that a significant fraction of their findings could not be replicated by subsequent investigations.

This might leave you with a rather bleak view of scientific research. Indeed, many in the community have been sounding the alarm about the lack of replicability of published results and have proposed some basic remedies. Perhaps the most important is the

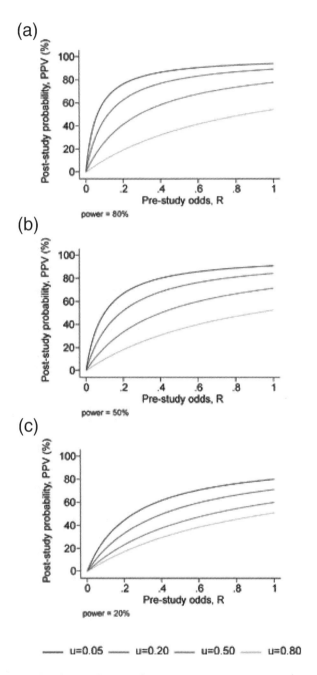

Figure 7.3: The dependence of post-study probability (positive predictive value) on the pre-study odds, for different power of the study, and with different levels of bias. Figure from Ioannidis (2005b) under CC-BY.

issue that only positive results are deemed worthy of publication. If all 20 groups in the scenario described above published their results, 19 would report no effect, 1 would report an effect, and the picture would be clear. There are some journals (e.g., *PLoS One*) that accept any methodologically sound submission, regardless of whether the result is positive or negative. Another remedy is to provide funding for research labs to repeat other groups' studies to test them. These steps are being implemented, and hopefully will eventually lead to an improvement in the reliability of published data. Even more importantly, educating scientists about basic probability ideas, such as Bayes' formula and the notion of prior knowledge, should improve the quality of inference and decrease the amount of questionable science.

Discussion questions:

The following questions refer to the paper "Why Most Published Research Findings Are False" (Ioannidis , 2005b).

Discussion 7.3.1. What are some of the examples of studies with low prior odds that Ioannidis uses as examples?

Discussion 7.3.2. Due to human imperfection, there will always be some bias in conducting and reporting scientific results. What do you expect to be the typical level of bias (designated u in the paper)?

Discussion 7.3.3. What does Ioannidis propose to remedy the problem of lack of reproducibility of studies? Will you take them into account when reading scientific publications or doing your own research?

7.4 Conditional statements and random simulations

Here we introduce *conditional statements*, which utilize `if` and `else` commands. The `if` command must be followed by a *logical test*: for instance, `a>b` or `a == b`; in the first case, the commands following the if statement (in brackets) are executed if the variable `a` is greater than b, and in the second case, they will be executed if a equals b. If the logical test is false, then the commands following the else statement (in brackets) are executed.

We can use a random number from the uniform random number generator to "draw" a patient from a population with a certain prevalence of the disease. For a given prevalence set by prob (e.g., 0.2), if the random number is less than prob, the patient will be counted as diseased, and otherwise as healthy. This is implemented in the following script:

```
prob <- 0.2   # set probability of disease
decider <- runif(1)   # random number between 0 and 1
Disease <- 0
Healthy <- 0
if (decider < prob) {
    Disease <- Disease + 1
    # increment number of diseased
} else {
    Healthy <- Healthy + 1
    # increment number of healthy
}
```

This script only produces one patient, so at the end either Disease is 1, or Healthy is 1, while the other one is zero. To generate a "sample" of patients, we can use a for loop to repeat this random decision multiple times. The following script repeats the process a specified number of times:

```
prob <- 0.2   # set probability of disease
Disease <- 0
Healthy <- 0
numpatients <- 100
for (i in 1:numpatients) {
    decider <- runif(1)   # random number between 0 and 1
    if (decider < prob) {
        Disease <- Disease + 1
        # increment number of diseased
```

7.4. RANDOM SIMULATIONS

```
    } else {
        Healthy <- Healthy + 1
        # increment number of healthy
    }
}
print(Disease)

## [1] 10

print(Healthy)

## [1] 90
```

This script produces a simulated sample of 100 patients out of a population with 20% disease prevalence; it then prints out how many are diseased and how many are healthy.

You can use nested conditional statements to simulate a number of hypothesis tests with a given sensitivity and specificity. The hypothesis has a prior probability of being true, which we will simulate by drawing a random number and assigning our individual experiment a value (e.g., 0 or 1). The following script uses two conditional statements to make two random decisions: is the hypothesis actually true or false (e.g., is the patient sick or healthy), and what the outcome of the test is (reject or not reject the hypothesis) based on the sensitivity and specificity. Since the second decision depends on the first, you will need to nest the conditional statements for the test inside the conditional statements for the hypothesis. The following script simulates testing 100 randomly selected patients from a population with 20% disease prevalence, and testing them with a test with 0.9 sensitivity (90% of truly sick people test positive) and 0.99 specificity (99% of truly healthy people test negative):

```
prob <- 0.2   # set probability of disease
TP <- 0    # true positive number
FP <- 0    # false positive number
TN <- 0    # true negative number
FN <- 0    # false negative number
```

```
sens <- 0.9   # sensitivity of the test
spec <- 0.99  # specificity of the test
numpatients <- 100
for (i in 1:numpatients) {
    decider <- runif(1)    # random number between 0 and 1
    decider2 <- runif(1)   #generate second random number
    if (decider < prob) {
        # the patient is sick
        if (decider2 < sens) {
            TP <- TP + 1   # increment the true positive
        } else {
            FN <- FN + 1   # increment the false negative
        }
    } else {
        # patient is healthy
        if (decider2 < spec) {
            TN <- TN + 1   # increment the true negative
        } else {
            FP <- FP + 1   # increment the false positive
        }
    }
}
print(paste("true positive", TP))

## [1] "true positive 17"

print(paste("true negative", TN))

## [1] "true negative 81"

print(paste("false positive", FP))

## [1] "false positive 0"

print(paste("false negative", FN))

## [1] "false negative 2"
```

7.4. RANDOM SIMULATIONS

Look carefully at the logical structure of the script, because it is easy to make a mistake. If the patient is sick, there are two possible test outcomes: a true positive or a false negative; if a patient is healthy, the two possibilities are a true negative or a false positive.

Programming exercises:

Find the errors in the scripts presented below by copying each script into an R editor window and debugging it until it runs and does what it is intended to do.

Exercise 7.4.1. This script is supposed to draw a random number between 0 and 1, and if it is less than 0.5, print "Heads!" or otherwise print "Tails!"

```
if (decider < 0.5) {
      decider <- runif(1)
      print('Heads!')
} else {}
      decider <- runif(1)
      print['Tails!']
```

Exercise 7.4.2. The following script is supposed to simulate a sample of 500 patients from a population with disease prevalence of 10%, and saving their disease status ("sick" or "healthy") in a vector.

```
numpatients <- 500 # set number of patients
prev <- 0.1 # set disease prevalence
for { (j in 1:numpatients)
  decider <- runif(1)
  if (decider < prev) { # a sick patient
    Status_Vector[i] <- 'sick'
  }
}
```

Exercise 7.4.3. This script is supposed to take the vector of disease status (assigned in the previous script); run a hypothesis test on each one, with a false positive rate of 0.01 and false negative rate of 0.1; and then record the results in a vector.

```
FPR <- 0.01 # set the false positive rate
FNR <- 0.1 # set the false negative rate
# pre-allocate vector of test results
test.results<-rep('blah',numpatients)
decider <- runif(1)
for (i in 1:numpatients) {
  if (Status_Vector=='sick') {
      if (decider < FNR) { # false negative result
          test.results <- 'negative'
      } else { # true positive result
          test.results <- 'positive'
      }
   }
   if (Status_Vector=='healthy') {
      if (decider < FPR) { # false positive result
          test.results <- 'positive'
      } else { # true negative result
          test.results <- 'negative'
      }
   }
}
```

7.5 Computational project

This project illustrates the ideas of the Ioannidis (2005b) paper described above. The basic idea is that if a hypothesis has a small prior probability of being true (e.g., looking through an entire genome for single nucleotide polymorphisms (SNPs) that are linked with a disease) then a positive result has a low predictive value. We will simulate this by controlling the prior probability of the hypothesis being true and the sensitivity and specificity of the test. The following script uses a random number to decide whether a particular SNP is linked to some disease, based on a prior probability.

7.5. COMPUTATIONAL PROJECT

```
prior.prob <- 0.2   # prob of SNP linked to disease
decider <- runif(1)   # random number between 0 and 1
if (decider < prior.prob) {
   # if random number < prob
   link <- 1   # SNP is linked to disease
} else {
   link <- 0   # SNP is independent of disease
}
```

To simulate the whole experiment, we need to make two separate random decisions:

1. Is the hypothesis actually true or false (e.g., is the randomly selected SNP linked to the disease)?

2. Is the test outcome positive (reject the hypothesis of no linkage) or negative (do not reject the hypothesis)?

Since the second decision depends on the first, you need to nest the conditional statements by putting the one determining the test outcome inside the conditional statements deciding whether the hypothesis is true. This is shown in the following script, which simulates running a hypothesis test for linkage of SNP and disease with given sensitivity and specificity.

```
spec <- 0.7   # set specificity
sens <- 0.8   # set sensitivity
TP <- 0
FP <- 0
TN <- 0
FN <- 0
decider <- runif(1)   # random number between 0 and 1
if (link == 1) {
    if (decider < sens) {
        # test correctly identifies the linkage
        TP <- TP + 1   # true positive result
    } else {
        FN <- FN + 1   # false negative result
```

```
        }
    } else {
        if (decider < spec) {
            # test correctly says there is no linkage
            TN <- TN + 1   # true negative result
        } else {
            FP <- FP + 1   # false positive result
        }
    }
}
print(paste("The number of true positives is", TP))
print(paste("The number of true negatives is", TN))
print(paste("The number of false positives is", FP))
print(paste("The number of false negatives is", FN))
```

Tasks

1. Use the second of the sample scripts provided above to simulate a test for a SNP that is linked to a disease (set `link` to 1) with specificity of 0.8 and sensitivity of 0.9. Put a for loop around the script to run it 100 times, and count how many of the generated test results are true positives, false positives, true negatives and false negatives. (Hint: only two of those are possible for a true hypothesis.) How many times does the hypothesis test make the correct decision?

2. Run the same script for a false hypothesis (set `link` to 0), with the same specificity and sensitivity of the test and count how many of the 100 test results are true positives, false positives, true negatives, and false negatives (hint: only two of those are possible for a false hypothesis.) How many times does the hypothesis test make the correct decision?

3. Use the first of the provided scripts to randomly simulate whether a particular SNP is linked with the prior probability 0.01 that the SNP is linked to the disease, and then follow with the second script to simulate the hypothesis test. Use the same sensitivity and specificity values as above, and run the loop for 1000 trials. Based on your counts of the different

7.5. COMPUTATIONAL PROJECT

test outcomes, report the positive predictive value of the test (the probability that the SNP is linked to disease, given that the test result is positive, or the fraction of true positives out of all positives).

4. Investigate the effect of changing the prior probability of the hypothesis being true. Change the prior probability to 0.1, use the same sensitivity and specificity values as before, and run the loop for 1000 trials and report the positive predictive value of the test. Now change the prior probability to 0.001, use the same sensitivity and specificity values as before, and run the loop for 1000 trials and report the positive predictive value of the test. How does the prior probability affect the positive predictive value? What implication does this have for testing a large number of hypotheses with low prior probabilities, such as thousands of SNPs in the human genome?

5. Investigate the effect of changing the sensitivity and specificity of the test. Set the prior probability to 0.1, set the sensitivity to be 0.9 and the specificity to be 0.9, run the loop for 1000 trials, and report the positive predictive value of the test. Now change the sensitivity to 0.99 and specificity to 0.9, run the loop for 1000 trials, and report the positive predictive value of the test. Finally, set the sensitivity to be 0.8 and the specificity to be 0.99, and again report the positive predictive value of the test. Which parameter (sensitivity or specificity) has the largest effect on the positive predictive value?

Chapter 8

Linear regression and correlation

> *The place in which I'll fit will not exist until I make it.*
> —James Baldwin

In Chapters 6 and 7 you learned to use data that fall into a few categories. We now turn to data that can be measured as a range of numerical values. We can ask a similar question of numerical data that we asked of categorical: How can we tell whether two variables are related? And if they are, what kind of relationship is it? This takes us into the realm of *data fitting*, raising two related questions: What is the best mathematical relationship to describe a data set? What is the quality of the fit? You will learn to do the following in this chapter:

1. measure the quality of the fit between a line and a two-variable data set,

2. calculate the parameters for the best-fit line based on statistics of the data set,

3. use R to calculate and plot the best-fit line for a data set,

4. understand the meaning correlation and covariance, and

5. understand the phenomenon of regression to the mean.

8.1 Linear relationship between two variables

Although real data always contain errors, there may be a relationship between the two variables that is not random: say, when one goes up, the other one tends to go up as well. These relationships may be complicated, so in this chapter we focus on the the simplest and most common type of relationship: linear, where a change in one variable is associated with a *proportional* change in the other plus an added constant. This is expressed mathematically using the familiar equation for a linear function with parameters slope (m) and intercept (b):

$$y = mx + b$$

Let us say you have measured some data for two variables, which we will call, unimaginatively, x and y. This data set consists of pairs of numbers: one for x, one for y; for example, the heart rate and body temperature of a person go together. They cannot be mixed up between different people, as the data will lose all meaning. We can denote this by a list of n pairs of numbers: (x_i, y_i), where i is an integer between 1 and n. Since this is a list of pairs of numbers, we can plot them as separate points in the plane using each x_i as the x-coordinate and each y_i as the y-coordinate. This is called a *scatterplot* of a two-variable data set. For example, two scatterplots of a data set of heart rate and body temperature are shown in Figure 8.1. In the first one, the body temperature is on the x-axis, which makes it the *explanatory* variable; in the second one, the body temperature is on the y-axis, which makes it the *response* variable.

8.2 Linear least-squares fitting

8.2.1 sum of squared errors

It is easy to find the best-fit line for a data set with only two points: its slope and intercept can be found by solving the two simultaneous linear equations. For example, if the data set consists of

8.2. LINEAR LEAST-SQUARES FITTING

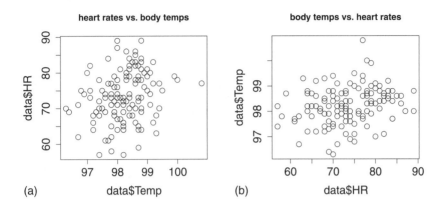

Figure 8.1: Scatterplots of heart rates and body temperatures: (a) with heart rate as the explanatory variable; (b) with body temperature as the explanatory variable.

$(3, 2.3), (6, 1.7)$, then finding the best fit values of m and b means solving the following two equations:

$$3m + b = 2.3$$
$$6m + b = 1.7$$

These equations have a unique solution for each unknown: $m = -0.2$ and $b = 2.9$ (you can solve it using basic algebra).

However, a data set with two points is too small and cannot serve as a reasonable guide for finding a relationship between two variables. Let us add one more data point to increase our sample size to three: $(3, 2.3), (6, 1.7), (9, 1.3)$. How do you find the best fit slope and intercept? **Bad idea:** take two points and find a line (i.e., the slope and the intercept) that passes through the two. It should be clear why this is a bad idea: we are arbitrarily ignoring some of the data, while perfectly fitting two points. So how do we use all the data? Let us write down the equations that a line with slope m and intercept b has to satisfy to fit our data points:

$$3m + b = 2.3$$
$$6m + b = 1.7$$
$$9m + b = 1.3$$

This system has no exact solution, since there are three equations and only two unknowns. We need to find m and b such that they are a "best fit" to the data, not the perfect solution. To do that, we need to define what we mean by the *goodness of fit*.

One simple way to assess how close the fit is to the data is to subtract the predicted values of y from the data as follows: $e_i = y_i - (mx_i + b)$. The values e_i are called the *errors* or *residuals* of the linear fit. If the values predicted by the linear model $(mx_i + b)$ are close to the actual data y_i, then the error will be small. However, if we add it all up, the errors with opposite signs will cancel each other, giving the impression of a good fit simply if the deviations are symmetric.

A more reasonable approach is to take absolute values of the deviations before adding them up. This is called the *total deviation* for n data points with a line fit:

$$TD = \sum_{i=1}^{n} |y_i - mx_i - b|$$

Mathematically, a better measure of total error is a sum of squared errors (SSE), which also has the advantage of adding up non-negative values but is known as a better measure of the distance between the fit and the data (think of Euclidean distance, which is also a sum of squares):

$$SSE = \sum_{i=1}^{n} (y_i - mx_i - b)^2$$

Thus we have formulated the goal of fitting the best line to a two-variable data set, also known as linear regression: find the values of slope and intercept that result in the lowest possible SSE. A mathematical recipe exists that produces these values, which is described in the next section. Any model begins with assumptions,

8.2. LINEAR LEAST-SQUARES FITTING

and for linear regression to be a faithful representation of a data set, the following must be true:

- the variables have a linear relationship,
- all measurements are independent of one another,
- there is no noise in the measurements of the explanatory variable, and
- the noise in the measurements of the response variable is normally distributed with mean 0 and identical standard deviation.

The reasons these assumptions are necessary for linear regression to work are beyond the scope of the text; they are well elucidated in *Numerical Recipes* (Press et al. 2007). However, it is important to be aware of them, because if they are violated, the resulting linear fit may be meaningless. It's fairly clear that if the first assumption is violated, you are trying to impose a linear relationship on something that is actually curvy. The second assumption of independence is very important and often overlooked. The mathematical reasons for it have to do with properly measuring the goodness of fit, but intuitively, it is because measurements that are linked can introduce a new relationship that has to do with the measurements rather than the relationship between the variables. Violation of this assumption can seriously damage the reliability of the linear regression. The third assumption is often ignored, since usually the explanatory variable is also measured and thus has some noise. The reason for it is that the measure of goodness of fit is based only on the response variable, and there is no consideration of the noise in the explanatory variable. However, a reasonable amount of noise in the explanatory variable is not catastrophic for linear regression. Finally, the last assumption is due to the statistics of maximum-likelihood estimation of the slope and intercept, but again, some deviation from perfect normality (bell-shaped distribution) of the noise, or slightly different variation in the noise, is to be expected.

8.2.2 best-fit slope and intercept

To calculate the linear relationship between two variables, we first need to measure how much of their variances is linked.

Definition 8.1. The *covariance* of a data set of pairs of values (X, Y) is the sum of the products of the corresponding deviations from their respective means:

$$Cov(X, Y) = \frac{1}{n-1} \sum_{i=1}^{n} (x_i - \bar{X})(y_i - \bar{Y})$$

Intuitively, this means that if two variables tend to deviate in the same direction from their respective means, they have a positive covariance, and if they tend to deviate in opposite directions from their means, they have a negative covariance. In the intermediate case, if sometimes they deviate together and other times they deviate in opposition, the covariance is small or zero. For instance, the covariance between two independent random variables is zero, as we saw in section 4.2.

It should come as no surprise that the slope of linear regression depends on the covariance, that is, the degree to which the two variables deviate together from their means. If the covariance is positive, then for larger values of x the corresponding y values tend to be larger, which means the slope of the line is positive. Conversely, if the covariance is negative, so is the slope of the line. And if the two variables are independent, the slope has to be close to zero. The actual formula for the slope of the linear regression is (Whitlock and Schluter 2008):

$$m = \frac{Cov(X, Y)}{Var(X)} \qquad (8.1)$$

I will not provide a proof that this slope generates the minimal sum of squared errors, but that is indeed the case. To find the intercept of the linear regression, we make use of one other property of the best fit line: for it to minimize the SSE, it must pass through the point $(\bar{X}, \bar{Y})$. Again, I will not prove this, but given that the point of the two mean values is the central point of the "cloud" of

8.2. LINEAR LEAST-SQUARES FITTING

points in the scatterplot, it should be plausible that if the line missed that central point, the deviations will be larger. Assuming that is the case, we have the following equation for the line: $\bar{Y} = m\bar{X} + b$, which we can solve for the intercept b:

$$b = \bar{Y} - \frac{Cov(X,Y)\bar{X}}{Var(X)} \tag{8.2}$$

Math exercises:

Use the data set in Table 8.1 to answer the following questions:

B (m^2/kg)	H ($°C/min$)
7.0	0.103
5.0	0.091
3.6	0.014
3.3	0.024
2.4	0.031
2.1	0.006

Table 8.1: Body leanness (B) and heat loss rate (H) in boys; partial data set from Sloan and Keatinge (1973).

Exercise 8.2.1. Compute the mean and standard deviation of each variable.

Exercise 8.2.2. Compute the covariance between the two variables.

Exercise 8.2.3. Calculate the slope and intercept of the linear regression for the data with B as the explanatory variable.

Exercise 8.2.4. Make a scatterplot of the data set with B as the explanatory variable, and sketch the linear regression line with the parameters you computed.

Exercise 8.2.5. Calculate the slope and intercept of the linear regression the data with H as the explanatory variable.

Exercise 8.2.6. Make a scatterplot of the data set, with H as the explanatory variable, and sketch the linear regression line with the parameters you computed.

8.2.3 correlation and goodness of fit

The correlation between two random variables is a measure of how much variation in one is linked to variation in the other. If this sounds similar to the description of covariance, it's because they are closely related. Essentially, correlation is "normalized" covariance, made to range between -1 and 1.

Definition 8.2. The *(linear or Pearson) correlation* of a dataset of pairs of data values (X, Y) is

$$r = \frac{Cov(X,Y)}{\sqrt{Var(X)Var(Y)}} = \frac{Cov(X,Y)}{\sigma_X \sigma_Y}$$

If the two variables are identical, $X = Y$, then the covariance becomes its variance $Cov(X,Y) = Var(X)$, the denominator also becomes the variance, and the correlation is 1. This is also true if X and Y are scalar multiples of each other, as you can see by plugging in $X = cY$ into the covariance formula. The opposite case is if X and Y are diametrically opposite, $X = -cY$, which has the correlation coefficient of -1. All other cases fall in the middle, neither perfect correlation nor perfect anticorrelation. In the special case if the two variables are independent, and thus their covariance is zero, the correlation coefficient of 0.

This definition connects correlation and the slope of linear regression:

$$m = r \frac{\sigma_Y}{\sigma_X} \tag{8.3}$$

Whenever linear regression is reported, one always sees the values of correlation r and squared correlation r^2 displayed. This is because r^2 has a clear meaning as the the fraction of the variance of the dependent variable Y explained by the linear regression $Y = mX + b$. Let us unpack what this means.

According to the stated assumptions of linear regression, the response variable Y is assumed to have a linear relationship with the

8.2. LINEAR LEAST-SQUARES FITTING

explanatory variable X, but with independent additive noise (also normally distributed, but it doesn't play a role in this argument). Linear regression captures the linear relationship, and the remaining errors (residuals) represent the noise. Thus, each value of Y can be written as $Y = R + \hat{Y}$, where R is the residual (noise) and the value predicted by the linear regression is $\hat{Y} = mX + b$. The assumption that R is independent of Y means that $Var(Y) = Var(\hat{Y}) + Var(R)$, because variance is additive for independent random variables, as discussed in section 4.2. By the same reasoning, $Cov(X, \hat{Y} + R) = Cov(X, \hat{Y}) + Cov(X, R)$. These two covariances can be simplified further: $Cov(X, R) = 0$ because R is independent random noise. The variable X and the predicted $\hat{Y}$ are perfectly correlated, so $Cov(X, \hat{Y}) = Cov(X, mX + b) = Var(X) = Var(\hat{Y})$. This leads to the derivation of the meaning of r^2:

$$r^2 = \frac{Cov(X,Y)^2}{Var(X)Var(Y)} = \frac{(Cov(X,\hat{Y}) + Cov(X,R))^2}{Var(X)Var(Y)} = \\ = \frac{Var(X)Var(\hat{Y})}{Var(X)Var(Y)} = \frac{Var(\hat{Y})}{Var(Y)} \tag{8.4}$$

One should be cautious when interpreting results of a linear regression. First, just because there is no linear relationship does not mean that there is no other relationship. Figure 8.2 shows some examples of scatterplots and their corresponding correlation coefficients. What it shows is that while a formless blob of a scatterplot will certainly have zero correlation, so will other scatterplots in which there is a definite relationship (e.g., a circle, or an X-shape). The point is that correlation is always a measure of the linear relationship between variables.

The second caution is well known: the danger of equating correlation with a causal relationship. There are numerous examples of scientists misinterpreting a coincidental correlation as meaningful, or deeming two variables that have a common source as causing each other. For example, one can look at the increase in automobile ownership over the past century and the concurrent improvement in longevity and conclude that automobiles are good for human health. It is well documented, however, that a sedentary lifestyle and automobile exhaust do not make a person healthy. Instead, increasing

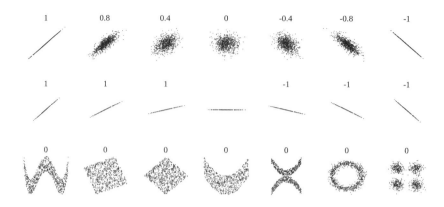

Figure 8.2: The correlation coefficient does not tell the whole story when it comes to describing the relationship between two variables. Images are from "Correlation examples2" by Imagecreator, updated by DenisBoigelot, in public domain via Wikimedia Commons.

prosperity has both enhanced the purchasing power of individuals and enabled advances in medicine that lengthens our lifespans. To summarize, one must be careful when interpreting correlation: a weak one does not mean there is no relationship, and a strong one does not mean that one variable causes the changes in the other.

Another important measure of the quality of linear regression is the *residual plot*. The residuals are the differences between the predicted values of the response variable and the actual value from the data. As stated above, linear regression assumes a linear relationship between the two variables, with some uncorrelated noise added to the values of the response variable. If that were true, then the plot of the residuals would look like a vaguely spherical blob, with a mean value of 0 and no discernible trend (e.g., no increase of residual for larger x values). An example can be seen in Figure 8.4 in section 8.3. Visually assessing residual plots is an essential check on whether linear regression is a reasonable fit to the data in addition to the r^2 value.

8.2. LINEAR LEAST-SQUARES FITTING

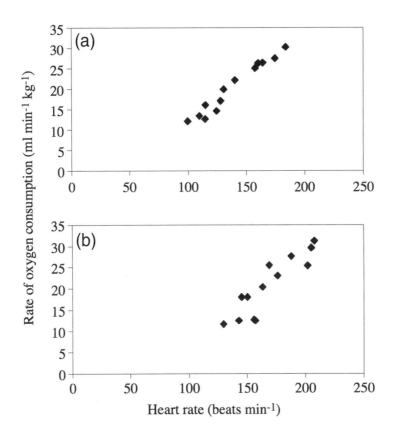

Figure 8.3: Mass-specific rate of oxygen consumption (VO) as a function of heart rate (HR) in two macaroni penguins, (a) a breeding female of mass 3.14 kg and (b) a moulting female of mass 3.99 kg. Figure from Green et al. (2001) under CC-BY.

Math exercises:

Figure 8.3 shows scatterplots of the rate of oxygen consumption (VO) and heart rate (HR) measured in two macaroni penguins running on a treadmill (really). The authors performed linear regression on the data and found the following parameters: $VO = 0.23 HR - 11.62$ (penguin A) and $VO = 0.25 HR - 20.93$ (penguin B). The data sets have the standard deviations $\sigma_{VO} = 6.77$ and $\sigma_{HR} = 28.8$ (penguin A), and $\sigma_{VO} = 8.49$ and $\sigma_{HR} = 30.6$ (penguin B).

Exercise 8.2.7. Find the dimensions and units of the slope and the intercept of the linear regression for this data (the units of HR and VO are given in Figure 9.3).

Exercise 8.2.8. Data set B has a larger slope than does data set A. Does this mean the correlation is higher in data set B than in A? Explain.

Exercise 8.2.9. Calculate the correlation coefficients for the linear regressions of the two penguins; explain how much variance is accounted for in each case.

Exercise 8.2.10. Recalculate the slopes of the two linear regressions if the explanatory and response variables were reversed. Does changing the order of variable affect the correlation?

8.3 Linear regression using R

We now have the tools to compute the parameters of the best-fit line, provided we can calculate the means, variances, and covariance of the two-variable data set. Of course, the best way to do all this is to let a computer handle it. The function for calculating linear regression in R is `lm()`, which outputs information to a variable called `myfit` in the script below. The slope, intercept, and other parameters can be printed out using the `summary()` function. In the script below prints out a bunch of information, but we are concerned with the values in the first column corresponding to the best-fit intercept (-166.2847) and the slope (2.4432). You can check that they correspond to our formulas by computing the covariance, the variances, and the means of the two variables.

```
data <- read.table("data/HR_temp.txt", header = TRUE)
myfit <- lm(HR ~ Temp, data)
summary(myfit)

##
## Call:
## lm(formula = HR ~ Temp, data = data)
```

8.3. LINEAR REGRESSION USING R

```
## 
## Residuals:
##     Min      1Q  Median      3Q     Max
## -16.6413 -4.6356  0.3247  4.8304 15.8474
## 
## Coefficients:
##              Estimate Std. Error t value Pr(>|t|)
## (Intercept) -166.2847    80.9123  -2.055  0.04190
## Temp           2.4432     0.8235   2.967  0.00359
## 
## (Intercept) *
## Temp        **
## ---
## Signif. codes:
## 0 '***' 0.001 '**' 0.01 '*' 0.05 '.' 0.1 ' ' 1
## 
## Residual standard error: 6.858
## on 128 degrees of freedom
## Multiple R-squared:  0.06434,
## Adjusted R-squared:  0.05703
## F-statistic: 8.802 on 1 and 128 DF, p-value: 0.003591

m <- cov(data$HR, data$Temp)/var(data$Temp)
print(m)

## [1] 2.443238

b <- mean(data$HR) - m * mean(data$Temp)
print(b)

## [1] -166.2847
```

Here `Temp` and `HR` are the explanatory and response variables, respectively, and `data` (optional) is the name of the data frame they are stored in. The best-fit parameters are stored in `myfit`, and the line can be plotted using `abline(myfit)`. The next script shows how to calculate a linear regression line and then plot it over a scatterplot in R. The result is shown in Figure 8.4a.

```
plot(data$Temp, data$HR, main = "scatterplot
    and linear regression line",
    cex = 1.5, cex.axis = 1.5, cex.lab = 1.5)
abline(myfit)
HRresiduals <- resid(myfit)
plot(data$Temp, HRresiduals, main = "residuals plot",
    cex = 1.5, cex.axis = 1.5, cex.lab = 1.5)
abline(0, 0)
```

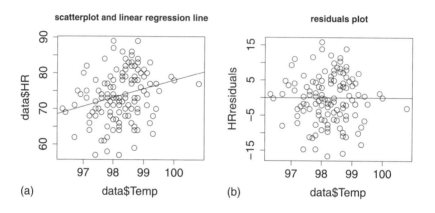

Figure 8.4: Linear regression for a data set of heart rates and body temperatures (a); and the residuals (b).

However, what does this mean about the quality of the fit? Just because we found a line to draw through a scatterplot does not mean that this line is meaningful. In fact, looking at the plot, there does not seem to be much of a relationship between the two variables. There are various statistical measures for the significance of linear regression; the most important one relies on the correlation between the two data sets. Look again at the summary statistics for the data set of heart rates and temperatures. Several different statistics are given here, and the one that we care about is the r^2, which is reported earlier in this section as "Multiple R-squared." This number tells us that the linear regression accounts for only about 6% of the total variance of the heart rate. In other words, there is no significant linear relationship in this data set.

8.4. REGRESSION TO THE MEAN

As mentioned in section 8.2, the other important check is plotting the residuals of the data set after the linear fit is subtracted. You see the result in right panel of figure 8.4, showing that the residuals do not have any pronounced pattern. So it is reasonable to conclude that linear regression was a reasonable model to which to fit the data. The low correlation is because the data seem to have little to no relationship, not because there is some complicated non-linear relationship.

8.4 Regression to the mean

The phenomenon called *regression to the mean* is initially surprising. Francis Galton first discovered this by comparing the heights of parents and their offspring. Galton took a subset of parents who are taller than average and observed that their children were, on average, shorter than their parents. He also compared the heights of parents who are shorter than average and found that their children were on average taller than their parents. This suggests the conclusion that in long run, everyone will converge closer to the average height—hence "regression to mediocrity," as Galton called it (Senn 2011).

But that is not the case! The parents and children in Galton's experiment had a very similar mean and standard deviation. This appears to be a paradox, but it is easily explained using linear regression. Consider two identically distributed random variables (X, Y) with a positive correlation r. The slope of the linear regression is $m = r\sigma_Y/\sigma_X$ and since $\sigma_Y = \sigma_X$, the slope is simply r. Select a subset with values of X higher than $\bar{X}$, and consider the mean value of Y for that subset. If the slope $m < 1$ (the correlation is not perfect), then the mean value of Y for that subset is less than the mean value of X. Similarly, for a subset with values of X lower than $\bar{X}$, the mean value of Y for that subset is greater than the mean value of X, again as long as the slope is less than 1.

Figure 8.5 shows Galton's data set (available in R by installing the package "HistData") along with the linear regression line and the identity line $(y = x)$. If each child had exactly the same height as the parents, the scatterplot would lie on the identity line. Instead,

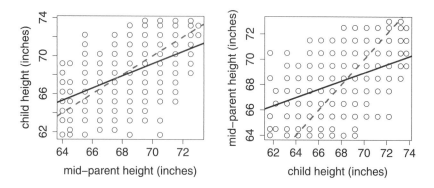

Figure 8.5: Galton's data on heights of parents and of children as scatterplots (two versions with explanatory and response variables switched). The dashed lines show the identity line $y = x$, and the solid lines are the linear regressions. For the plot on the left, the best-fit slope is about 0.65, for the plot on the right, the slope is about 0.33.

the linear regression lines have slope less than 1 for both the plot with the parental heights as the explanatory variable (left panel in Figure 8.5) and for the plot with the variables reversed (right panel). The correlation coefficient r does not depend on the order of the variables; so using equation 8.3, we can see the difference in slopes is explained by the two data sets having different standard deviations, and reversing the explanatory and response variables results in reciprocation of the ratio of standard deviations. The children's heights have a higher standard deviation, which is likely an artifact of the experiment. In the data set the heights of the two parents were averaged to take them both into account, which substantially reduces the spread between male and female heights. To summarize, although the children of taller parents are shorter on average than their parents, and the children of shorter parents are taller than their parents, the overall standard deviation does not decrease from generation to generation.

Discussion questions:

Read the paper on measuring the rate of de novo mutations and its relationship to paternal age (Kong et al., 2012).

Discussion 8.4.1. What types of mutations were observed in the data set? What were the most and the least common?

Discussion 8.4.2. The paper shows that both maternal and paternal age are positively correlated with offspring inheriting new mutations. What biological mechanism explains why paternal age is the dominant factor? What could explain the substantial correlation with maternal age?

Discussion 8.4.3. Is linear regression the best representation of the relationship between paternal age and number of mutations? What other model did the authors use to fit the data, and how did it perform?

Discussion 8.4.4. What do you make of the historical data of paternal ages the authors present at the end of the paper? Can you postulate a testable hypothesis based on this observation?

8.5 Computational projects

Here is a sample script for calculating and plotting linear regression in R:

```
x.data <- sample(0:100, 20)   # generate explanatory data
y.data <- -5 * x.data + 100 + 20 * runif(20)
# generate response
# linear regression with y.data as the response
# variable
myfit <- lm(y.data ~ x.data)
plot(x.data, y.data, xlab = "explanatory",
     ylab = "response")
abline(myfit)
summary(myfit)
```

Here `x.data` and `y.data` are the explanatory and response variables, respectively; one can add the name of the data frame in which they are stored, as an option. The best-fit parameters are assigned to `myfit`, and the line can be plotted using `abline(myfit)`. The information from the linear regression can be accessed by using `summary(myfit)`.

8.5.1 parental age and new mutations

In the following tasks you will analyze data from the paper by Kong et al. (2012). The data files can be downloaded from the textbook website: https://dkon.uchicago.edu/page/quantifying-life. The authors measured the number of new (de novo) mutations in humans by comparing whole-genome sequences of two parents and their biological children. One of the questions they addressed was the relationship between the number of new mutations and the ages of the two parents. The variables `PatAge` and `MatAge` contain the paternal and maternal ages, respectively (in years), and the variable `Mutations` contains the number of de novo mutations.

Tasks

1. Compute and plot the linear regression for `Mutations` as a function of `PatAge`. How does the slope compare to the fit your performed by eye in the computational project in Chapter 3? What does the slope of that line mean? What fraction of the variance is explained by the linear relationship?

2. Compute and plot the linear regression for `Mutations` as a function of `MatAge`. How does the slope compare to the fit your performed by eye in the project in Chapter 3? What does the slope of that line mean? What fraction of the variance is explained by the linear relationship? Is paternal or maternal age more strongly correlated with new mutations in the children?

3. Load the data file 3gen_mutation_data.txt, which contains the data for 5 families for which Kong et al. sequenced three generations of people and were able to distinguish mutations

8.5. COMPUTATIONAL PROJECTS

occurring in the paternal and maternal lineages. The data file contains variables PatAge and MatAge (self-explanatory), and PatMut and MatMut (number of mutations on paternal and maternal chromosomes, respectively). Perform linear regression between number of mutations in paternal lineage and paternal age, and between the number of mutations in maternal lineage and maternal age. Plot the results, and report the slope of the line and the fraction of variance explained by the linear relationship.

4. Using the same data set as in task 3, perform linear regression for paternal mutations and maternal age, and the maternal mutations and paternal age, plot the results, and report the slope and the fraction of variance explained by the linear relationship. If there is a substantial correlation, explain its meaning (e.g., does the age of the father affect the number of mutations inherited from the mother, and if so, how?)

8.5.2 heart rates on two different days

For this project you will use a data set generated by your class, which hopefully consists of at least a dozen students. Have every student measure their resting heart rate one day, and then repeat the measurement a day (or two) later. Call these variables HR1 and HR2, and have everyone record these values in a shared document (one can use a randomly generated personal code to make the data anonymous). This will produce a data set with two variables (heart rate 1 and heart rate 2), which should presumably be similar for each individual.

Tasks

1. Import the data set into R and perform linear regression on the two variables, with HR1 as the explanatory variable and HR2 as the response. Report the slope, the intercept, and the goodness of fit. How close is the scatterplot to the identity line?

2. Separate the subset of data points into two groups: those with below-mean heart rates on day 1 from those with above-mean heart rates. The easy way to do this is to use the `which()` function, which returns the indices of elements satisfying a certain condition. Report the mean heart rate for the first group on day 1 and on day 2 and compare them. Report the mean heart rate for the second group on day 1 and on day 2 and compare them.

3. Explain your observations in words. Why did the mean of group 1 increase and the mean of group 2 decrease from day 1 to day 2? How did I know in advance that the data would behave like that without ever seeing it?

Chapter 9

Nonlinear data fitting

Bend corners like I was a curve, I struck a nerve
And now you 'bout to see this Southern playa serve.
—OutKast, *ATLiens*

9.1 Nonlinear relationships between variables

Not all variables have the courtesy of relating to each other in a linear fashion. Some can be sneaky, and display a linear relationship over a range of values but then follow a different pattern over another range. Take, for example, the relationship between the size of a genome of an organism (in terms of the number of nucleotide base pairs) and the number of genes that code for proteins. The two variables are plotted in Figure 9.1 for 623 different bacterial and archeal genomes. The relationship is strongly linear and suggests a universal rule that larger genomes contain more protein-coding genes.

However, if ones adds genome data from eukaryotic organisms, the trend does not fit this prediction. Figure 9.2 shows data from 55 eukaryotic and 1055 non-eukaryotic genomes (including small genomes from viruses and organelles) plotted on a logarithmic scale. A linear relationship is still linear with slope 1 after being log-transformed, as we will derive in the next section, and the prokaryotic genomes produce the best-fit line with slope of 0.997. Hover,

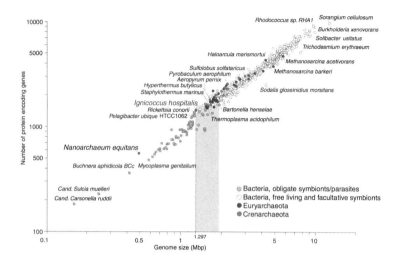

Figure 9.1: Relationship between genome size and number of protein coding genes for prokaryotes; Mbp = millions of base pairs. Figure from Podar et al. (2008) under CC-BY.

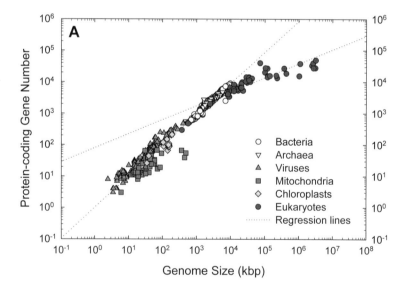

Figure 9.2: Relationship between genome size and number of protein coding genes for different life forms; kbp=thousands of base pairs. Figure from Hou and Lin (2009) under CC-BY.

the eukaryotic ones show a very different relationship, with number of genes increasing much more slowly for the same increase in genome size. The figure shows a different fit for this portion of the data, with the best-fit line (of the transformed variables) having the slope of 0.47.

In the chapter you will learn to do the following:

1. perform logarithmic transformation on data;

2. interpret the results of linear regression on log-transformed data; and

3. use R to fit nonlinear data, using both linear regression on log-transformed variables and polynomial fitting.

9.2 Nonlinear fitting using logarithmic transforms

Many biological processes exhibit an exponential dependence. For instance, biochemical reactions are frequently governed by exponential decay over time, and the rates of reactions are found by fitting the data to exponential function. The functional form of a exponential is $y = ae^{kx}$, where y the dependent variable, x is the independent variable, k is the exponential rate, and a is the multiplicative constant. One can take the logarithm of the dependent variable and of the right-hand side to produce the following transformed equation:

$$\ln(y) = \ln(ae^{kx}) = \ln(a) + kx$$

This shows that the logarithm of y has a linear relationship with x. Thus, one can use the linear least-squares fit between the log-transformed variable y and the original variable x to find the slope k, which corresponds to the exponential growth parameter, and the intercept, which is the logarithm of the constant multiplier parameter a. When the variable $\ln(y)$ is plotted against x, it is called a *semi-log* plot.

Other variables have another kind of relationship called a *power law*. The functional form appears to be similar to an exponential one, with one important distinction: the independent variable is the base and is not in the exponent of the expression. The general formula for a power law is: $y = ax^n$, where y is the dependent variable, x is the independent variable, n is the power parameter, and a is the multiplicative constant. Once again, we can transform this expression by taking the logarithm (of any base) of both sides:

$$\log(y) = \log(ax^n) = \log(a) + n\log(x)$$

The transformation demonstrates that the logarithm of y has a linear relationship with the logarithm of x. To obtain the values of the power law parameter, one can perform the linear least-squares fit between $\log(y)$ and $\log(x)$. The slope of this regression is the power parameter n, and the intercept is the logarithm of the multiplicative constant a. When the variable $\log(y)$ is plotted against $\log(x)$, it is called a *log-log* plot.

While logarithmic transforms followed by linear regression fitting are fairly common in science, one should be cautious when using them. The problem is that the logarithm changes the scale of the variables in a nonlinear fashion (as you can see on any logarithmic plot), and so modifies the magnitude of the noise that is part of any experimental measurement. Logarithms make large numbers disproportionally small, or equivalently, they magnify small numbers. One of the assumptions of linear regression is that the noise in the dependent variable must have equal variance across all the data points. By log-transforming the dependent variable in either of the two transformations mentioned above, the noise is magnified for the small values of y and is diminished for the large values. Depending on the variance in the noise in the original data, this may result in a highly disproportionate effect of the smaller values on the fit as opposed to the larger ones, which you will demonstrate for yourself in the computational project. In short, while this is a convenient and commonly used technique, it often produces dubious results, with exponential fits tending to underestimate the true exponential rates. A better approach is to use a fully nonlinear fitting algorithm, such as the Levenberg-Marquardt method (Press et al. 2007), but this technique is outside the scope of this introductory textbook.

9.3 Generalized linear fitting in R

9.3.1 logarithmic transforms

We saw in section 9.2 that we can transform certain functions from nonlinear to linear by using logarithmic transforms. Let us show the example of transforming a data based on the power law $y = 0.3x^{1.7}$ with added noise using the following script.

```
a <- 0.3
n <- 1.7
x.data <- seq(1, 10, 0.1)
y.data <- a * x.data^n + 2 * runif(length(x.data))
myfit <- lm(log(y.data) ~ log(x.data))
nf <- myfit$coefficients[2]
af <- exp(myfit$coefficients[1])
summary(myfit)

##
## Call:
## lm(formula = log(y.data) ~ log(x.data))
##
## Residuals:
##      Min       1Q   Median       3Q      Max
## -0.51530 -0.09047  0.01316  0.08480  0.75195
##
## Coefficients:
##             Estimate Std. Error t value Pr(>|t|)
## (Intercept) -0.31985    0.05775  -5.538 3.06e-07
## log(x.data)  1.29721    0.03468  37.401  < 2e-16
##
## (Intercept) ***
## log(x.data) ***
## ---
## Signif. codes:
## 0 '***' 0.001 '**' 0.01 '*' 0.05 '.' 0.1 ' ' 1
##
```

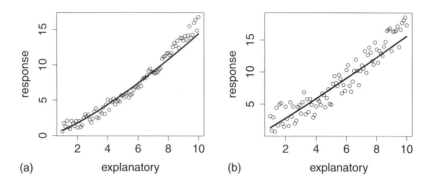

Figure 9.3: Two examples of fitting using a power law for a simulated data set with $y = 0.3x^{1.7}$ and with uniform added noise levels 2(a) and 5(b).

```
## Residual standard error: 0.1981
## on 89 degrees of freedom
## Multiple R-squared:  0.9402,
## Adjusted R-squared:  0.9395
## F-statistic:  1399 on 1 and 89 DF, p-value: < 2.2e-16
```

To plot the results, we take the coefficients in myfit and use them as parameters a and n for the power law function. The coefficients are stored in the output of the lm() function, which is a data frame, and can be accessed from the vector myfit$coeffficents; the first element is the intercept, and the second is the slope. This script produces a plot of the best-fit power law curve for the data set over the plot of the original, not log-transformed, variables. It is evident from the plots in Figure 9.3 that this approach systematically underestimates the exponent in the power law. This illustrates the limitations of using log-transformed variables in conjunction with linear regression.

9.3.2 polynomial regression

Some data sets that show nonlinear dependence may be fitted with a polynomial curve, for example, a parabola. The relationship has the

9.3. GENERALIZED LINEAR FITTING IN R

functional form $y = ax^2 + bx + c$, where y is the dependent variable; x is the independent variable; and a, b, and c are parameters to be fit. Similar to the problem of fitting a linear function to a two-variable data set discussed in section 8.1, this results in a set of n equations, one for each data point:

$$\begin{aligned} ax_1^2 + bx_1 + c &= y_1 \\ ax_2^2 + bx_2 + c &= y_2 \\ &\cdots \\ ax_n^2 + bx_n + c &= y_n \end{aligned}$$

From these equations we can write down the expression for the sum of squared errors and set the goal of minimizing it. As with linear regression, I am going to skip over the mathematical details of finding the optimal parameters a, b, and c. However, it is worth noting that although the form of the relationship is nonlinear, the polynomial fitting problem is linear because of the linear relationship between the sum of of squared errors and the parameters. Thus, the problem is actually called *generalized linear fitting*, and it can be done using the same R function lm() that we used for linear regression. Given two data variables x.data and y.data, the following script generates a set of data that is quadratic with known parameter values a, b, and c with random noise added. The script then performs a quadratic fit on the data and prints out the three best-fit parameters using the summary() function. The data together with the best-fit parabola are plotted in Figure 9.4.

```
a <- 3
b <- -25
c <- 10
x.data <- seq(1, 10, 0.1)
y.data <- a * x.data^2 + b * x.data + c + 10
        * runif(length(x.data))
myfit <- lm(y.data ~ poly(x.data, 2, raw = TRUE))
af <- myfit$coefficients[3]
bf <- myfit$coefficients[2]
cf <- myfit$coefficients[1]
summary(myfit)
```

```
## 
## Call:
## lm(formula = y.data ~ poly(x.data, 2, raw = TRUE))
## 
## Residuals:
##     Min      1Q  Median      3Q     Max
## -5.5898 -2.3959 -0.3706  2.1773  4.9995
## 
## Coefficients:
##                              Estimate Std. Error
## (Intercept)                  15.89915    1.30277
## poly(x.data, 2, raw = TRUE)1 -25.31984    0.53500
## poly(x.data, 2, raw = TRUE)2   3.01450    0.04756
##                              t value Pr(>|t|)
## (Intercept)                    12.20   <2e-16 ***
## poly(x.data, 2, raw = TRUE)1  -47.33   <2e-16 ***
## poly(x.data, 2, raw = TRUE)2   63.38   <2e-16 ***
## ---
## Signif. codes:
## 0 '***' 0.001 '**' 0.01 '*' 0.05 '.' 0.1 ' ' 1
## 
## Residual standard error: 2.8
## on 88 degrees of freedom
## Multiple R-squared:  0.9903,
## Adjusted R-squared:  0.99
## F-statistic:  4470 on 2 and 88 DF, p-value: < 2.2e-16
```

The function `poly()` generates a polynomial of a given degree; in this case it is quadratic. If the data are not fit well by a quadratic curve, one can try fitting with higher degree polynomials. However, there is an important question about the appropriate number of parameters in a fit for a particular data set. Adding more parameters always results in a better-looking fit, but at some point the number of parameters is too large, and *overfitting* becomes an issue. In fact, if one uses the same number of parameters as data points, one can obtain a perfect fit that has little predictive power—it just matches

9.4. ALLOMETRY AND POWER LAW SCALING

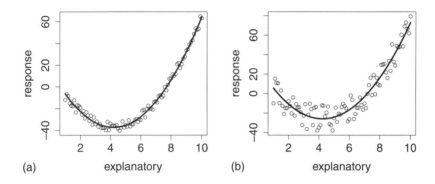

Figure 9.4: Two examples of a quadratic fit for a simulated data set with $y = 3x^2 - 25x + 10$ and with uniform added noise levels 10(a) and 30(b).

the given data. Deciding at what point adding more parameters is not productive is a difficult question; it can be addressed by various statistical methods that are outside of the scope of this text.

9.4 Allometry and power law scaling

The relationship between the size of an animal and its various physiological properties is the subject of the field of *allometry*. For example, the width of bones of land-dwelling animals is not proportional to their size. The bones of an elephant are proportionately much larger and thicker than those of a mouse. One interesting relationship that has been studied is the one between the size (mass) of animals and their basal metabolic rate (rate of energy consumption at rest). In the 1930s Kleiber plotted a log-log plot of the two variables for about a dozen different animals (West and Brown 2005), ranging is size from a dove to a steer. The plot is shown in Figure 9.5, and it looks remarkably linear with slope close to 3/4. As discussed in section 9.2, this corresponds to the power law relationship with power 3/4, which is a curve that grows a little bit slower than a straight line (which has power 1).

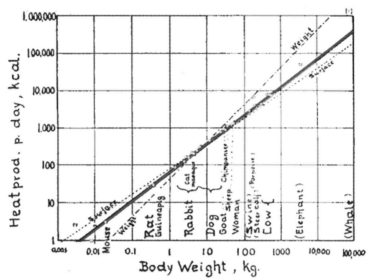

Figure 9.5: Log-log plot of the basal metabolic rate and mass for 13 warm-blooded animals. Figure by Max Kleiber in public domain via Wikimedia commons.

The origin of this power law and the significance of the power 3/4 has fascinated theoretical biologists ever since. Ingenious models have been proposed, based on the principle of self-similarity, that result in the prediction of the mysterious power of 3/4. However, as data on animals and their metabolism accumulated, it became apparent that the power of 3/4 does not fit all the data equally well. If one limits the data to animals of smaller size, a smaller power is the result of a fit, but for larger animals, the power is larger. Recently, Kolokotrones and others took a different approach and fit the log-transformed data with a quadratic fit (Kolokotrones et al., 2010). A parabola can be thought of as line with a changing slope, so this in principle could match the observation. Figure 9.6 shows the scatterplot of the data after double logarithmic transform, along with the best-fit line and the best-fit parabola. The quadratic fit not only does a better job, judging by the fitting statistics, it also more accurately matches the correct basal metabolic rates of two

9.4. ALLOMETRY AND POWER LAW SCALING

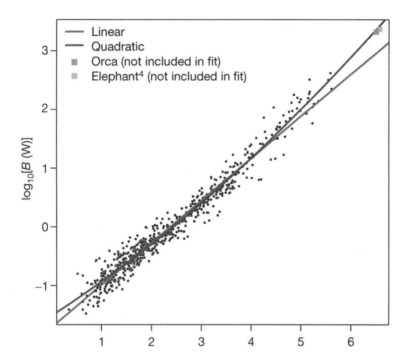

Figure 9.6: Log-log plot of the basal metabolic rate and mass for many warm-blooded animals, with best-fit linear and quadratic curves overlaid. Figure from Kolokotrones et al. (2010) used by permission.

large animals (orca and elephant) that were not included in the fit. This suggests that the relationship between the two variables is not a power law but a more complicated relationship captured by the quadratic function between the logarithms of the variables.

Discussion questions:

The following questions refer to the paper "Curvature in Metabolic Scaling" (Kolokotrones et al., 2010).

Discussion 9.4.1. What are some limitations inherent in studying the relationship between measurements taken from related species? How do the authors address them?

Discussion 9.4.2. What explanation do the authors give for the changing slope of the relationship between mass and metabolism? Do you find them plausible?

Discussion 9.4.3. An earlier theory based on vascular structure was used to derive the 3/4 power law relationship. How do the authors contend with that earlier model, and what do they propose instead?

Discussion 9.4.4. While the quadratic fit is empirically better at representing the data, do you find it more intellectually satisfying than the simple power law model of Kleiber?

9.5 Computational projects

In this project you will fit a data set with an exponential relationship by transforming the exponential function into a linear form and using linear regression to fit the parameters. To do this, take the logarithm of the dependent variable:

$$y = B + Ae^{kx} \Rightarrow y - B = Ae^{kx} \Rightarrow \log(y - B) = \log(A) + kx$$

As you can see, after subtracting the response variable y from B (the asymptotic value) and taking the logarithm, the resulting variable has a linear relationship with the explanatory variable x. So if you know the asymptote B (which you can easily measure from the graph), you can use linear regression to find $\log(A)$ (the intercept) and k (the slope) from the linear regression. Here is a sample script to use.

```
x.data <- runif(20)
# generate some explanatory data points
A <- -5
B <- 100
k <- -4
# generate response data points with an
# exponential, plus noise
y.data <- A * exp(k * x.data) + B + 0.1 * runif(20)
# fit the log-transformed data to a linear
```

9.5. COMPUTATIONAL PROJECTS

```
# relationship
myfit <- lm(log(abs(y.data - B)) ~ x.data)
plot(x.data, y.data)
summary(myfit)
```

To plot the results, take the coefficients from `myfit`, assign them to parameters A and k, as was done in section 9.3, and use them with the exponential function to plot your fit using the `curve()` function.

Load the data files Insulin_data.txt and KaiC_data.txt, and fit the time dependence of the data sets using an exponential function, as follows. The data files can be downloaded from the textbook website: https://dkon.uchicago.edu/page/quantifying-life.

Tasks

1. Perform linear regression on the data set of insulin concentration after log-transformation, and plot the graph of the fit using lines over the scatterplot. Compare the result with the parameters you obtained by estimating the parameters in the project in Chapter 4.

2. You might see that the fit is not very good, because the logarithmic transform gives too much weight to the slow-changing phase of the data. Instead, restrict the data set to the first 10 time points of the insulin data, obtain new parameters from linear regression (choose a different asymptote B), and produce a new graph. Does it fit better? How does its rate compare to the one you estimated by eye?

3. Perform linear regression on the data set of KaiC after log-transformation, and plot the graph of the fit using lines over the scatterplot. Compare the result with the parameters you obtained by estimating the parameters in the project in Chapter 4.

4. Once again, to improve the fit, restrict the data set to the first 8 time points of the KaiC data, obtain new parameters from linear regression, and produce a new graph. Does it fit better? How does its rate compare to the one you estimated by eye?

Part III

Chains of random variables

Chapter 10

Markov models with discrete states

> *True, man is mortal, but that's not the half of it. What's worse is that he's sometimes suddenly mortal, that's the trick!*
>
> —Mikhail Bulgakov, *The Master and Margarita*

Life is complex and often unpredictable. Molecules bump into each randomly due to thermal motion; entire organisms either find food or become food themselves due to chance. Mathematical probability supplies tools to model, analyze, and even predict the behavior of these random processes. In this part of the book we focus on living systems that can be described as being in different categories called *discrete states*. Similar to the categorical random variables that were described in Chapters 6 and 7, these systems are not measured on a numerical scale but can be described by words. For example, *ion channels* are transmembrane proteins that can change their shape to allow or not allow the passage of ions. Thus, an ion channel can be described as being in an open or a closed state, with transitions taking places between those states with certain probabilities. These models with a few discrete states and random transitions with specified probabilities are called *Markov models*. They are easy to build, and they provide a powerful framework for mathematical analysis. In this chapter you will learn to do the following:

1. write down the transition diagram and transition matrix of a discrete-state Markov model,

2. understand the Markov property and its implications,

3. calculate the probability of a string of states based on transition probabilities, and

4. simulate a Markov model by generating multiple state strings based on transition probabilities.

10.1 Building Markov models

Consider the life cycle of a cell, which is illustrated in Figure 10.1a. Cells are known to go through phases in the *cell cycle*, which correspond to different molecules being synthesized and different actions being performed. The M phase stands for mitosis, or cell division, which itself can be divided into stages. In between cell divisions cells go through gap phases (G_1 and G_2) and the S phase, during which DNA replication occurs.

The three non-mitotic phases are sometimes grouped into Interphase (I). Some cells, depending on their environmental conditions or their type in multicellular organisms, can also get off the treadmill of the cell cycle and go into what is called "quiescence," or the G_0 phase, during which the cell leads a quiet life. It can also come out of quiescence and replicate again. This suggests a simplified description of the cell that exists either in state R (actively replicating) or state Q (quiescent), with transitions between the two states occurring randomly with some probabilities. These kinds of models can be summarized graphically using *transition diagrams*. For example, the QR model of the cell cycle with probability of transition from Q to R of 0.05 (per hour), and the transition probability from R to Q of 0.1 (per hour) is shown in Figure 10.1b.

A very different example of biological systems naturally divided into states are ion channels, mentioned above. For some of them the opening or closing is activated by the binding of other molecules, as in the case of the nicotinic acetylcholine receptor (nAChR). When

10.1. BUILDING MARKOV MODELS

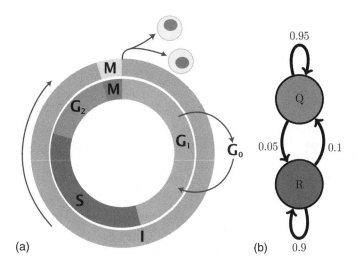

Figure 10.1: (a) Diagram of the cell cycle showing the replicating phases (M, G_1, S, and G_2) and the quiescent phase G_0. Image "Cell Cycle" by Zephyris with modifications by Beao and Histidine under CC-BY-SA-3.0 via Wikimedia Commons. (b) Diagram of a model of the cell cycle with two states (Q and R) with transitions between states shown as arrows labeled with transition probabilities.

not bound to acetylcholine (a small molecule that serves as a neurotransmitter) it remains closed, but binding acetylcholine enables it to change conformation and open, though it can also be closed when bound to acetylcholine. Figure 10.2 illustrates the three states of nAChR, along with a transition diagram that depicts possible transitions. Notice the absence of any arrows between states R and O, which reflects the fact that the ion channel cannot transition directly from the resting (unbound) state to the open state; it must go through the bound-but-closed state C.

10.2 Markov property

In these *discrete-time finite-state* Markov models the dependent variable is the state of the object (e.g., cell, ion channel), and the independent variable (e.g., time) advances in discrete steps, the length of which is defined by the problem. For example, in the cell cycle model an appropriate time step may be an hour, while for the ion channel mode a reasonable time step is a fraction of a second. In some bioinformatics models describing a string of letters the independent variable is position in the sequence, and the step is one letter. Changes from one state to another are called *transitions*, and they may only happen over a step of the independent variable. The transitions occur randomly, so they cannot be predicted, but we can describe the probability of transitions.

For example, we may state that the probability of transition from state Q to R in the cell cycle model is 0.05 for each time step, and the probability of transition from R to Q is 0.1. This means that 5 times out of 100 (out of many trials) a quiescent cell will switch to replicating over one time step, and 1 time out of 10 (out of many trials) a replicating cell will switch to the quiescent state. Let us define these parameters properly.

Definition 10.1. Let $X(t)$ be the random variable in a discrete-time Markov model with finitely many states at an arbitrary time t. The *transition probability* from state i to state j is denoted p_{ji} and is defined as the conditional probability:

$$p_{ji} = P(X(t+1) = j | X(t) = i)$$

10.2. MARKOV PROPERTY

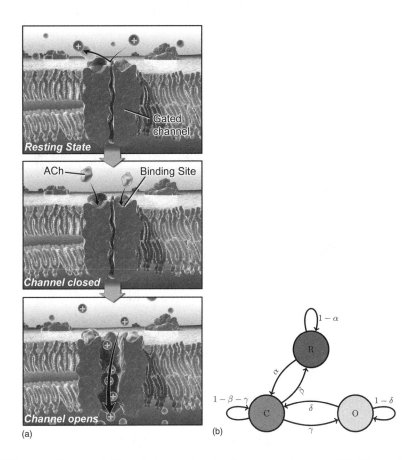

Figure 10.2: (a) Nicotinic acetylcholine receptor (nAChR) an ion channel that opens only when bound to acetylcholine (ACh); conformation of the ion channel can be divided into three states: resting (R), closed with ACh bound (C), and open (O). Image "Chemically Gated Channel" by Blausen.com under CC BY 3.0 via Wikimedia Commons. (b) Transition diagram illustrates the possible transitions with transition probabilities as parameters.

Let us unpack this definition. The transition probability is conditional on knowing the state of the model (i) at the present time (t) and gives us the probability of the model switching to another state (j, which can be the same as i) one time step later ($t+1$). The transition probability in this definition has no explicit dependence on time t, which is not always true—I just chose to make that additional assumption, called *time homogeneity*. There are Markov models which are not time homogeneous, but we will not discuss them in this textbook.

Note that the transition from state i to state j is written as p_{ji}. This convention is used to conform to the conditional probability notation, and for another reason that will be apparent in Chapter 11. Unfortunately, there is no agreement in the field as to which convention to use, and some textbooks and papers denote the same transition probability p_{ij}. To avoid confusion, I will remind you what p_{ji} stands for.

One funny thing that you might have noticed in the definition is that the transition probability makes no mention of times before the present. We just brazenly assumed that the history of the random variable before the present time t does not matter! This is called the *Markov property* and was first postulated by A. A. Markov in 1905. Here is the proper definition (Feller 1968).

Definition 10.2. A time-dependent random variable $X(t)$ has the *Markov property* if for all times t and for all $n < t$, the following is true:
$$P(X(t+1)|X(t); X(n)) = P(X(t+1)|X(t))$$

I did not specify the state of the random variable in the definition because it must be true for all states in the model. Stated in words, this says that the probability distribution of the random variable at the next time step, given its distribution at the current time, is the same whether any of its past states are known or not. Another way of stating it is that the state of the random variable at the next time, given the state at the present time, is independent of the past states.

If this seems like a really big assumption, you are right! The reason we assume this property is because it makes calculations

10.2. MARKOV PROPERTY

with these models much easier. As with any assumption, it must be viewed critically for any given application. Does the cell really forget what state it was in an hour ago? Does whether an ion channel was open a microsecond ago affect its probability to open again? The answer to these questions is not always clear cut— there is almost always some residual memory of past states in a real system. If that memory is not very strong, then we can proceed with our Markov modeling. Otherwise, the models must be made more sophisticated.

10.2.1 transition matrices

Let us return to the example of the cell cycle model with two transition probabilities given: transition from Q to R $p_{RQ} = 0.05$ and transition from R to Q $p_{QR} = 0.1$. We can calculate the probability of a replicating cell remaining in the same state and the probability of a quiescent cell remaining in the same state, because they are complementary events:

$$p_{QQ} = 1 - p_{RQ} = 0.95;\ p_{RR} = 1 - p_{QR} = 0.9$$

In other words, a quiescent cell either becomes replicating over one time step or remains quiescent, so the two probabilities must add up to 1. The same reasoning applies to the replicating cell. We now have all the parameters of the model, and there is a convenient way of organizing them in one object.

Definition 10.3. The *transition matrix* for a discrete-time Markov model with N states is an N-by-N matrix, which has the transition probabilities p_{ij} as its elements in the ith row and jth column.

By convention, the rows in matrices are counted from top to bottom, while the columns are counted left to right. The transition matrix is a square matrix consisting of all transition probabilities of a given Markov model. It is, in essence, what defines a Markov model, because the transition probabilities are its parameters.

Example. For the cell cycle model in Figure 10.1, the transition matrix is

$$M = \begin{pmatrix} 0.95 & 0.1 \\ 0.05 & 0.9 \end{pmatrix}$$

To write it down we have to put the states in order; in this case I chose Q to be state number 1, and R to be state number 2. This is entirely arbitrary, but must be specified for the matrix to have meaning. Notice that the probabilities of staying in a state are on the *diagonal* of the matrix, where the row and the column number are the same. The probability of transition from state 1 (Q) to state 2 (R) is in column 1, row 2, and the probability of transition from state 2 (R) to state 1 (Q) is in column 2, row 1.

Example. For the three-state model of the nAChR ion channel in Figure 10.2 the transition matrix is:

$$M = \begin{pmatrix} 1-\alpha & \beta & 0 \\ \alpha & 1-\beta-\gamma & \delta \\ 0 & \gamma & 1-\delta \end{pmatrix}$$

The matrix is for states R, C, and O, placed in that order. Notice that the transition probability between R and O and vice versa is zero, in accordance with the transition diagram. The probability of remaining in each state is 1 minus the sum of all the transition probabilities of exiting that state; for example, for state 2 (C) the probability of remaining is $1-\beta-\gamma$.

10.2.2 probability of a string of states

Knowing the parameters of the model gives us the tools to make probabilistic calculations. The simplest task is to find the probability of occurrence of a given string of states. For instance, for the cell cycle model, suppose we know that a cell is initially quiescent. What is the probability that it remains quiescent for two hours? The probability of the cell remaining quiescent for one time step is $p_{QQ} = 0.95$, and the probability of it remaining quiescent for one more time step is also p_{QQ}. Due to the Markov property, the two transitions are independent of each other, so the probabilities can be multiplied (due to the multiplicative property of independent events) to give the answer: $P\{QQQ\} = 0.95 \times 0.95 = 0.9025$.

The magic of Markov property allows us to calculate the probability of any string of states, given the initial state, as the product of transition probabilities. We can write this formally as follows.

10.2. MARKOV PROPERTY

For a string of states $S = \{x_1, x_2, x_3, ..., x_{T-1}, x_T\}$, where x_t represents the state at time t, the probability of this string, given that $P(x_1) = 1$, is

$$P(S) = p_{x_2 x_1} p_{x_3 x_2} \cdots p_{x_T x_{T-1}} = p_{x_T x_{T-1}} \cdots p_{x_3 x_2} p_{x_2 x_1}$$

The ellipsis represents all the intermediate transitions from state x_3 to state x_{T-1}. I also showed that reversing the order of multiplication, which is allowed because of commutativity, makes the order of states proceed more clearly.

Example. Let us calculate the probability of another string of states based on the the cell cycle model in Figure 10.1. The probability that a cell is initially in state R, then transitions to state Q and remains in state Q is a product of the transition probability from R to Q and the transition probability from Q to Q:

$$P(\{RQQ\}) = 0.1 \times 0.95 = 0.095$$

Notice that there is no transition probability for the first state; it simply must be specified.

Example. Let us calculate the probability of a string of states based on the three-state model of the nAChR ion channel in Figure 10.2. The probability that an ion channel is initially in state R, remains in that state for 5 steps, then transitions to state C, remains there for 3 steps, and then transitions to state O is

$$P\{RRRRRRCCCCO\} = (1-\alpha)^5 \alpha (1-\beta-\gamma)^3 \gamma$$

Math exercises:

For the following Markov models (a) draw the transition diagram, if one is not provided; (b) put the states in (some) order, and write down the transition matrix; and (c) calculate the probability of the given strings of states, taking the first state as given (e.g., for a string of 3 states, there are only 2 transitions).

Exercise 10.2.1. Use the model in the transition diagram in Figure 10.3a to calculate the probability of the string of states BAB.

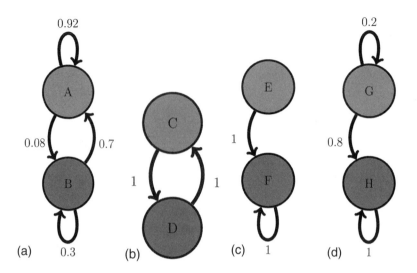

Figure 10.3: Transition diagrams for two-state Markov models.

Exercise 10.2.2. Use the model in the transition diagram in Figure 10.3b to calculate the probability of the string of states CCD.

Exercise 10.2.3. Use the model in the transition diagram in Figure 10.3c to calculate the probability of the string of states EEF.

Exercise 10.2.4. Use the model in the transition diagram in Figure 10.3d to calculate the probability of the string of states GGG.

Exercise 10.2.5. An ion channel can be in either open (O) or closed (C) states. If it is open, then it has probability 0.1 of closing in 1 microsecond; if closed, it has probability 0.3 of opening in 1 microsecond. Calculate the probability of the ion channel going through the following sequence of states: COO.

Exercise 10.2.6. An individual can be either susceptible (S) or infected (I), the probability of infection for a susceptible person is 0.05 per day, and the probability an infected person becoming susceptible is 0.12 per day. Calculate the probability of a person going through the following string of states: SISI.

Exercise 10.2.7. The genotype of an organism can be either normal (wild type, W) or mutant (M). Each generation, a wild type

10.2. MARKOV PROPERTY

individual has probability 0.03 of having a mutant offspring, and a mutant has probability 0.005 of having a wild type offspring. Calculate the probability of a string of the following genotypes in successive generations: WWWW.

Exercise 10.2.8. There are three kinds of vegetation in an ecosystem: grass (G), shrubs (S), and trees (T) (Bodine, Lenhart, and Gross 2014). Every year 25% of grassland plots are converted to shrubs, 20% of shrub plots are converted to trees, 8% of trees are converted to shrubs, and 1% of trees are converted to grass; the other transition probabilities are 0. Calculate the probability of a plot of land having the following succession of vegetation from year to year: GSGG.

Exercise 10.2.9. The nAChR ion channel can be in one of three states: resting (R), closed with ACh bound (C), and open (O) with transition probabilities (per microsecond): 0.04 (from R to C), 0.07 (from C to R), 0.12 (from C to O), and 0.02 (from O to C); the other transition probabilities are 0. Calculate the probability of the following string of states: OCCR.

Thinking problems:

Problem 10.2.1. We considered a sequence of Bernoulli trials in Chapter 4, for example, a string of coin tosses where each time heads and tails come up with probability 0.5. Describe this experiment as a Markov model, draw its transition diagram, and write its transition matrix.

Problem 10.2.2. Repeat problem 10.2.1 for a sequence of Bernoulli trials where success has probability 0.9 (and failure has probability 0.1).

Problem 10.2.3. Can you formulate a criterion based on a transition matrix of a Markov model, to tell whether it is generating a string of independent random variables as opposed to a string of random variables that depend on the previous one?

10.3 Simulation of Markov models

The behavior of a Markov model is random, so the future states cannot be predicted. However, one can use a computer to simulate the behavior of a Markov model, given its transition probabilities and an initial value. This is a useful approach to building intuition about the behavior of Markov models, because one can generate many strings of states, or "histories" of the model, and observe how much time the system spends in each state. This is what is called a *computer simulation*, which does not give exact results but is instead a numerical experiment, producing data that are consistent with a given model and its assumptions.

Here is a code to simulate one time step for a two-state Markov model. It generates a new state according to two transition probabilities, given an initial state. In the code I set the initial state to 1, and the transition probabilities are 0.6 and 0.4. The code uses a conditional statement to check what the initial state (`in.state`) is, and based on this use either transition probability from 1 to 2 (`trans1to2`) or transition probability from 2 to 1 (`trans2to1`) with a random number to make a random transition. A second conditional statement is used to assign `new_state` to a new state if the random number is less than the transition probability, and otherwise to leave `new_state` the same as `in.state`. The code also prints an error message if the initial state is neither 1 nor 2; it's good practice to cover all possibilities and not assume that variables are set to the values that you expect. If this code is run multiple times, you will see that the transitions are random: sometimes the model remains in the same state, and other times it jumps to the other.

```
in.state <- 1   # set initial state
trans1to2 <- 0.6   # transition probability from 1 to 2
trans2to1 <- 0.4   # transition probability from 2 to 1
decider <- runif(1)   # random num between 0 and 1
if (in.state == 1) {
    if (decider < trans1to2) {
        # randomly decide to transition
        new_state <- 2
```

10.3. SIMULATION OF MARKOV MODELS

```
    } else {
        # or to stay
        new_state <- 1
    }
} else if (in.state == 2) {
    if (decider < trans2to1) {
        # randomly decide to transition
        new_state <- 1
    } else {
        # or to stay
        new_state <- 2
    }
} else {
    print("Initial state must be either 1 or 2!")
}
print(new_state)

## [1] 2
```

This code only simulates a single transition. Let's simulate transitions over many time steps and generate a whole string of states! You could of course copy and paste the code above a bunch of times, but that would be ridiculous, not only because of the length of the resulting code, but also because each new state variable would need a new name. The smart way to generate a string of states is to use the tools of repetitive computation: a for loop and a vector variable. The code below uses the for loop to repeat the same random transition for as many steps as the programmer wishes, by setting the variable nsteps, and stores the resulting string of states (as 1s and 2s) in a vector variable states, which is plotted at the end. I run the simulation again and plot the two strings of states in Figure 10.4.

```
in.state <- 1  # set initial state
nsteps <- 50   # number of steps
# initialize the state.vec vector
state.vec <- rep(in.state, nsteps + 1)
```

```
trans1to2 <- 0.6   # transition probability from 1 to 2
trans2to1 <- 0.4   # transition probability from 2 to 1
for (i in 1:nsteps) {
    decider <- runif(1)   # generate random number
    if (state.vec[i] == 1) {
        if (decider < trans1to2) {
            # randomly decide to transition
            state.vec[i + 1] <- 2
        } else {
            # or to stay
            state.vec[i + 1] <- 1
        }
    } else if (state.vec[i] == 2) {
        # randomly decide to transition
        if (decider < trans2to1) {
            state.vec[i + 1] <- 1
        } else {
            # or to stay
            state.vec[i + 1] <- 2
        }
    } else {
        print("All states must be either 1 or 2!")
    }
}
```

10.4 Markov models of medical treatment

Markov models are used in many fields of biology. One example is the representation of disease and patient treatment using discrete states with random transitions. The states may describe the progression of the disease, the prior health and socioeconomic status of the patient, the treatment the patient is undergoing, or anything that is relevant for the medical situation. Some models are simple, for example, a model of stroke patients that describes them either as well, experiencing stroke, disabled, or dead (Sonnenberg and Beck 1993). Others use tens or hundreds of states, to better capture all

10.4. MARKOV MODELS OF MEDICAL TREATMENT

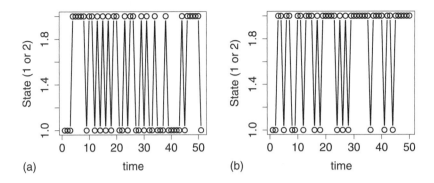

Figure 10.4: Examples of simulated state strings for a two-state model with transition probabilities (a) 0.4 and (b) 0.6, starting in state 1.

the details. One can then ask the question: What course of treatment is likely to lead to the best outcome? Notice that this question has no absolutely certain answer, since the model is fundamentally random. To evaluate different treatments, one can run simulations of the different models and compare the statistics generated by multiple simulations of each model.

A more realistic example comes from a recent study of the cost-effectiveness of different treatment protocols for HIV patients in South Africa (Leisegang et al., 2013). Physicians and public health officials have to consider both the costs and the efficacy of medical procedures, and it is a difficult challenge to balance the two, with human health and lives at stake. The authors built a model that includes stages of the disease, determined by viral loads and treatment options; the transition diagram is shown in Figure 10.5. The authors used data from two different clinics, a public and a private one, to estimate the transition probabilities and outcomes of treatments, and the costs of clinic visits and the therapies. They then simulated the models to compare the predicted costs and outcomes, and found that while the outcomes in the two treatment protocols were similar, the costs in the private clinics were considerably lower.

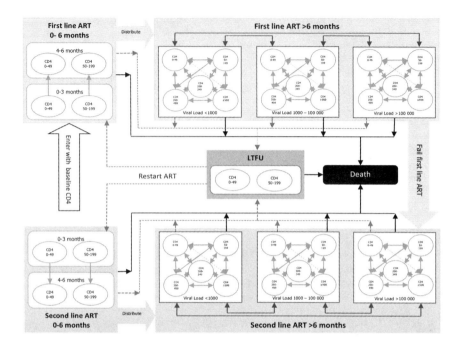

Figure 10.5: Transition diagram for a model of HIV disease and treatment. From Leisegang et al. (2013) under CC-BY.

Discussion questions:

The following questions refer to the paper (Leisegang et al., 2013) on modeling HIV treatment

Discussion 10.4.1. What does the Markov property mean for this model? How realistic do you think it is for actual patients?

Discussion 10.4.2. What are some other assumptions the authors make? Are they reasonable from a medical standpoint?

Discussion 10.4.3. The model predicts that private clinics that cut costs by reducing clinic visits are as effective as public clinics with greater numbers of visits. Would you be comfortable recommending patients use only private clinics based on this prediction?

10.5 Computational projects

10.5.1 state strings for a two-state model

In this section you will simulate a two-state Markov model by starting with a value of the state variable, using a random number and an if statement to generate the next value, and then the next one, and so on. Since Markov models are inherently random, each simulation will result in a different sequence of states.

Consider a simple model of an infectious epidemic, in which the population consists of only two types of people: susceptible (S) and infected (I). Suppose that the transition probabilities are (per day): 0.04 (from S to I) and 0.12 (from I to S). The first task asks you to write a script to decide, based on the initial state of the person (S or I), what state they are in the next day. The second task asks you to repeat that decision (using a for loop) to generate a string of states (S and I) over a given number of days.

Tasks

1. Write a script to a make a random decision for an individual based on the SI Markov model, where you set the initial state to be either 1 (S) or 2 (I). The script should first use an if statement to check whether the state is 1 or else is 2, and then use the transition probabilities given above to make the random decision. Run this script a few times to make sure it transitions from both state 1 to 2, and from 2 to 1 (remember that it's a random decision, so don't worry if the transition doesn't happen as often as you'd expect), and report the results of your experiment.

2. Add a for loop around your previous script to generate a state string over 100 days for an individual given an an initial state. Change your state variable into a vector (e.g., `state.vec`), and save the states for each day (labeled with numbers 1 and 2 for S and I). Plot that vector variable using `plot(state.vec)` to visualize the simulated disease history of one person, and plot the histogram of state frequencies using the command `barplot(table(state.vec))`. Rerun the script five times to

see different disease histories, and report the fraction of days these "people" spend infected.

3. Change the infection rate (transition from S to I) to 0.2, and repeat the computations in the previous task. Rerun the script five times to see different disease histories, and report the fraction of days these "people" spend infected. How different is the fraction in this case, compared to the previous task?

10.5.2 state strings for a three-state model

In this project you will construct and analyze a simple model of nAChR, which is an important ion channel involved in the transmission of signals between neurons. The opening of the channel is induced by binding of the neurotransmitter ACh. The Markov chain model has three states: resting (R), closed with ACh bound (C), and open (O). Suppose that at a certain concentration of Ach, the transition probabilities between the different states are as follows (per microsecond): 0.04 (from R to C), 0.07 (from C to R), 0.12 (from C to O) and 0.02 (from O to C); the other transition probabilities are 0.

Tasks

1. Write a script to a make a random decision for a single ion channel based on the Markov model above, where you set the initial state to 1 (C) (because it can transition to either R or O). The script should use an if statement followed by else if to check whether the state is 1 (R), 2 (C), or 3 (O); then use the transition probabilities given above to make the random decision. Run this script a few times to make sure it transitions from C to both R and O, and report the results of your experimentation.

2. Add a for loop around your previous script to generate a state string over 1000 microseconds for an ion channel starting with state R (1). Change your state variable into a vector, and save the states for each day (labeled with numbers 1, 2, and 3 for the states) into a vector (e.g., `state.vec`). Plot that vector

10.5. COMPUTATIONAL PROJECTS

variable using `plot(state.vec)` to visualize the simulated history of an ion channel, and plot the histogram of state frequencies using the command `barplot(table(state.vec))`. Rerun the script five times to see different ion channel state histories and report the fraction of time the ion channels are open.

3. Change the opening rate (transition from C to O) to 0.02, and repeat the computations in the previous task, Rerun the script five times to see different ion channel state histories, and report the fraction of time these ion channels are open. Compare the fractions to the results in the previous task, and describe the effect of changing the transition probability on the fraction of time the channels are open.

Chapter 11

Probability distributions of Markov chains

> *I had the most rare of feelings, the sense that the world, so consistently overwhelming and incomprehensible, in fact had an order, oblique as it may seem, and I a place within it.*
> —Nicole Krauss, *Great House*

In Chapter 10 you learned to describe randomly changing biological systems using discrete-state Markov models. These models are defined by a list of states and the transition probabilities between the states, which are organized into a transition matrix. The models are fundamentally stochastic and thus unpredictable, but they can be simulated on a computer using random number generators to produce multiple strings of states. From them one can calculate various statistical properties, such as means or variances of random variables that depend on these states. However, performing endless simulations can be computationally expensive. It is much more efficient to predict the probability distribution of the model at any given time without performing random simulations. The mathematical framework of matrices and vectors and algebraic operations on them allow this prediction. Here is what you will learn to do in this chapter:

1. write down a probability distribution vector for a Markov model;

2. given a probability distribution vector at a particular time, calculate the probability distribution one (or a few) time steps into the future;

3. multiply matrices and vectors; and

4. use R scripts to calculate the probability distribution for any number of time steps in the future.

11.1 Probability distributions evolve over time

Consider a two-state Markov model with a given transition matrix and a specified initial state. After one time step, the variable can be in either of two states—this can be simulated in R using a random number generator and a conditional statement. There are only two possible options: either the variable stays in the original state, or it transitions from the original state to the other one. For the cell cycle model introduced in Chapter 10, with initial state Q, the state space of this experiment contains two state strings: {QQ, QR}. After two time steps, there are now four possible paths. For the cell cycle model with initial state Q, the state space is {QQQ, QRQ, QRR, QQR}. Based on the calculation of probabilities of a given state discussed in Chapter 10, we can find the probability of each of the paths and then calculate the probability of the cell being in the replicating state after two time steps.

Let us make the problem a bit more general: take a two-state Markov model with a given transition matrix and a specified initial probability distribution, instead of a single state. In the cell cycle model, let the initial probability of Q be Q_0 and the initial probability of R be R_0. Then the event of the cell being in state Q after 1 time step is a combination of two different state strings: {QQ, RQ} and the probability of that event is the sum of those two probabilities:

$$P(X(1) = Q) = p_{QR}R_0 + p_{QQ}Q_0$$

Similarly, the event of the cell being in state R is a sum of the probabilities of the state strings {QR, RR}:

$$P(X(1) = R) = p_{RQ}Q_0 + p_{RR}R_0$$

11.1. DISTRIBUTIONS EVOLVE OVER TIME

These calculations are shown in Figure 11.1, starting with all the probability in state Q at $t = 0$, then calculating the new probability distribution at time $t = 1$, then using the same transition probabilities to calculate the new distribution at $t = 2$.

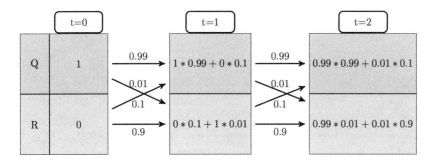

Figure 11.1: Transition probabilities used to calculate the evolution of probability in the QR model for two time steps.

The calculations are easy for a couple of time steps, but imagine having to do this for ten time steps, or a hundred—you would need to deal with thousands, or in the second case, about 10^{30} different state strings! One may use R to run a simulations for many different state strings and plot their histograms (which we will do in section 12.5). However, it is far more efficient to compute the probability of a cell being in a particular state at a particular time, that is, its *probability distribution vector*.

11.1.1 Markov chains

While sometimes the simulation of multiple random processes is necessary, it can get expensive. Probability offers us a theoretical way to make predictions for a Markov model, based on the notion of a *probability distribution vector*. First, let us define a few terms: a *Markov chain* is the mathematical manifestation of a Markov model. It was dubbed a "chain," because it consists of a string of linked random variables, the probability of each dependent on the previous one and generating the next one.

Definition 11.1. A *Markov chain* is a sequence of random variables $X(t)$, where t is the independent variable (e.g., time), and $X(t)$ can be in one of n states. Each random variable $X(t)$ has an associated probability distribution vector $\vec{P}(t)$, in which the ith element contains the probability of the random variable being in the ith state, and $\vec{P}(t)$ depends on $\vec{P}(t-1)$ according to the Markov property.

Example. The model of ion channels introduced in Figure 10.2 has three states: resting (R), closed (C), and open (O). A Markov chain for this model consists of a string of random variables that can take on states R, C, O, which can be indexed by integers 1, 2, 3, respectively. For each time step t, the random variable $X(t)$ has a probability vector with three elements: $\vec{P}(t) = (P_1(t), P_2(t), P_3(t))$. Each element represents the probability of the corresponding state at the time; for example, $P_3(20)$ is the probability of the ion channel being in state 3 (*O*) at time 20.

To generate a Markov chain with their associated probability distribution vectors, one needs to know the initial state or distribution. For example, if initially the ion channel model is in state C, the probability of the ion channel being in state R after 2 time step is different than if the initial state were R. Given an initial probability distribution, we will calculate the probability distribution at the next time step and then recursively generate the entire Markov chain.

The law of total probability (section 7.2) allows us to calculate the probability distribution of the Markov random variable at any time step, given its probability distribution at the previous time step. Here is the general formula for a Markov model with N states:

$$P(X(t+1) = i) = \sum_{j=1}^{N} P(X(t+1) = i | X(t) = j) P(X(t) = j)$$

$$= \sum_{j=1}^{N} p_{ij} P(X(t) = j) \qquad (11.1)$$

Here $P(X(t) = j)$ is the probability of the random variable being in some state j at time t, p_{ij} are the transition probabilities, and the

11.1. DISTRIBUTIONS EVOLVE OVER TIME

sum adds up all the possible transitions into state i. This equation can be written down for every state i at the next time step $t+1$, so we have N equations, each one adding up N terms. For a 2-state model, this is not too daunting:

$$P(X(t+1) = Q) = p_{QQ}P(X(t) = Q) + p_{QR}P(X(t) = R)$$

$$P(X(t+1) = R) = p_{RQ}P(X(t) = Q) + p_{RR}P(X(t) = R)$$

This gives us a predictive formula for the probability distribution of a Markov random variable at the next time point, given its current probability distribution. Notice that all four transition probabilities are used in the system of equations, and that they are arranged in the same way that I defined them in the transition matrix. This leads to a great simplification using matrices and vectors.

11.1.2 matrix multiplication

Now is a good time to properly define what matrices are and how we can operate on them. We have already seen transition matrices, but I introduce the following definitions just to make sure all the terms are clear.

Definition 11.2. A *matrix* A is a rectangular array of *elements* A_{ij}, in which i denotes the row number (index), counted from top to bottom, and j denotes the column number (index), counted from left to right. The *dimensions* of a matrix are defined by the number of rows and columns, so an *n-by-m matrix* contains n rows and m columns.

Elements of a matrix are not all created equal. They are divided into two types.

Definition 11.3. The elements of a matrix A that have the same row and column index (e.g., A_{33}) are called the *diagonal elements*. Those that do not lie on the diagonal are called the *off-diagonal elements*.

For instance, in the 3-by-3 matrix below, the elements a, e, and i are the diagonal elements:

$$A = \begin{pmatrix} a & b & c \\ d & e & f \\ g & h & i \end{pmatrix}$$

Matrices can be added together if they have the same dimensions. Then matrix addition is defined simply as adding up corresponding elements; for instance, the element in the second row and first column of matrix A is added to the element in the second row and first column of matrix B to give the element in the second row and first column of matrix C. Recall from Chapter 10 that rows in matrices are counted from top to bottom, while the columns are counted left to right.

Matrices can also be multiplied, but this operation is trickier. For mathematical reasons, multiplication of matrices $A \times B$ does not mean multiplying corresponding elements. Instead, the definition seeks to capture the calculation of simultaneous equations, like in the example with 2-state model in section 11.1.1. Here is the definition of matrix multiplication in words and in a formula (Strang 2005).

Definition 11.4. The *product of matrices A and B* is defined to be a matrix C, whose element c_{ij} is the dot product of the ith row of A and the jth column of B:

$$c_{ij} = a_{i1}b_{1j} + a_{i2}b_{2j} + \cdots + a_{iN}b_{Nj} = \sum_{k=1}^{q} a_{ik}b_{kj}$$

This definition is possible only if the length of the rows of A and the length of columns of B are the same, since we cannot compute the dot product of two vectors of different lengths. Matrix multiplication is defined only if A is n by q and B is q by m, for any integers n, q, and m, and the resulting matrix C is a matrix with n rows and m columns, as shown in Figure 11.2. In other words, the inner dimensions (number of columns of the matrix on the left and the number of rows of the matrix on the right) of matrices have to match for matrix multiplication to be possible.

11.1. DISTRIBUTIONS EVOLVE OVER TIME

Example. Let us multiply two matrices to illustrate how it's done. Here both matrices are 2 by 2, so their inner dimensions match and the resulting matrix is 2 by 2 as well:

$$\begin{pmatrix} 1 & 3 \\ 6 & 1 \end{pmatrix} \times \begin{pmatrix} 4 & 1 \\ 5 & -1 \end{pmatrix} = \begin{pmatrix} 1 \times 4 + 3 \times 5 & 1 \times 1 + 3 \times (-1) \\ 6 \times 4 + 1 \times 5 & 6 \times 1 + 1 \times (-1) \end{pmatrix}$$

$$= \begin{pmatrix} 19 & -2 \\ 29 & 5 \end{pmatrix}$$

One important consequence of this definition is that **matrix multiplication is not commutative**. If you switch the order, (e.g., $B \times A$), the resulting multiplication requires dot products of the rows of B by the columns of A, and except in special circumstances, they are not the same. In fact, unless m and n are the same integer, the product of $B \times A$ may not be defined at all.

We usually think of vectors as an ordered collection of numbers, but they can also be thought of as matrices, albeit skinny ones. A *column vector* is a matrix that has only one column, and a *row vector* is a matrix with only one row. Even if a column vector and a row vector contain the same numbers in the same order, they are different matrices, because they function differently when multiplied. When multiplying an n-by-m matrix and a vector, one can multiply with the vector on the left or on the right, depending on the type of vector: if it is a column vector, it must be on the right side of the matrix, while a row vector is multiplied on the left. Since the inner dimensions have to match, an n-by-m matrix can be multiplied by a m-by-1 column vector on the right, or by a 1-by-n row vector on the left.

The rules of matrix multiplication may seem annoyingly baroque, but you will see the payoff in the simplification of our Markov calculations.

Math exercises:

For the following pairs of matrices determine whether matrix multiplication is valid for $A \times B$ and $B \times A$, and for the valid cases, indicate the dimension of the resulting matrix.

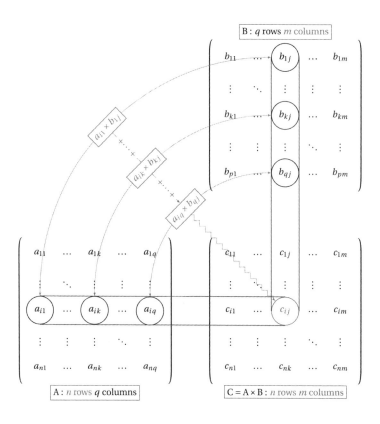

Figure 11.2: Matrix multiplication illustrating the details of calculating the matrix product $A \times B = C$. The dot product of the ith row of matrix A and the jth column of matrix B produces the element c_{ij} in the ith row and the jth column of C. Based on figure by Alain Matthes via http://texample.net under CC-BY 2.5.

11.1. DISTRIBUTIONS EVOLVE OVER TIME

Exercise 11.1.1.
$$A = \begin{pmatrix} 1 \\ 3 \end{pmatrix} ; B = \begin{pmatrix} 4 & 1 \\ 5 & 1 \\ 6 & 1 \end{pmatrix}$$

Exercise 11.1.2.
$$A = \begin{pmatrix} 1 & -2 & 1 \\ -2 & 1 & -2 \\ 1 & -2 & 1 \end{pmatrix} ; B = \begin{pmatrix} 4 & 5 & 6 \\ -6 & -5 & -4 \end{pmatrix}$$

Exercise 11.1.3.
$$A = \begin{pmatrix} 2 & -1 \\ -3 & 1 \end{pmatrix} ; B = \begin{pmatrix} -1 & -1 \\ -3 & -2 \end{pmatrix}$$

Exercise 11.1.4.
$$A = \begin{pmatrix} 1 & -2 & 1 \\ -2 & 1 & -2 \\ 1 & -2 & 1 \end{pmatrix} ; B = \begin{pmatrix} -1 & -1 \\ -3 & -2 \end{pmatrix}$$

Exercise 11.1.5.
$$A = \begin{pmatrix} 1 \\ 4 \\ -2 \end{pmatrix} ; B = \begin{pmatrix} -1 & -1 & 10 \\ -3 & -2 & 0 \\ 0 & -1 & -7 \end{pmatrix}$$

Exercise 11.1.6.
$$A = \begin{pmatrix} -1 & 2 & -9 \end{pmatrix} ; B = \begin{pmatrix} -4 & -8 \\ 5 & 2 \\ -6 & 10 \end{pmatrix}$$

11.1.3 propagation of probability vectors

To calculate the probability of states in the future, we need to start with an initial probability distribution—let us call it $P(0)$. To advance it by one time step, multiply it by the transition matrix and obtain the probability distribution at $P(1)$. To calculate the probability distribution after two time steps, multiply the distribution vector $P(1)$ by the transition matrix and obtain the vector $P(2)$.

Example. Take the case of the cell cycle model, where the cell is initially in the quiescent state, the vector propagation looks like

$$P(1) = M \times P(0) = \begin{pmatrix} 0.95 & 0.1 \\ 0.05 & 0.9 \end{pmatrix} \begin{pmatrix} 1 \\ 0 \end{pmatrix} = \begin{pmatrix} 0.95 \\ 0.05 \end{pmatrix}$$

The Markov property allows us to calculate the probability distribution at the next step $(t+1)$ given the distribution at the current step (t). This means to find the distribution of the cell cycle model after two time steps, we multiply the present distribution vector by the matrix M:

$$P(2) = M \times P(1) = \begin{pmatrix} 0.95 & 0.1 \\ 0.05 & 0.9 \end{pmatrix} \begin{pmatrix} 0.95 \\ 0.05 \end{pmatrix} = \begin{pmatrix} 0.9075 \\ 0.0925 \end{pmatrix}$$

One can calculate the probability distribution vectors for as many time steps as needed by repeatedly multiplying the current probability distribution vector by the transition matrix. The general formula for the probability distribution of a Markov chain at any time t is

$$P(t) = M \times P(t-1) = M^t \times P(0) \qquad (11.2)$$

which may be expressed in terms of repeated matrix multiplications (or the matrix M raised to the power t) of the initial distribution vector $P(0)$. This shows that to predict the distribution in the future, we need to know only two things: the initial distribution and the transition matrix of the Markov model.

Math exercises:

Use the transition matrices you constructed for these models in Chapter 10 to calculate the probability distribution for two time steps into the future.

Exercise 11.1.7. Use the model in the transition diagram in Figure 10.3a. If initially the model is in state A, what is the probability it is in state B after 1 time step? After 2 time steps?

Exercise 11.1.8. Use the model in the transition diagram in Figure 10.3b. If initially the model is in state C, what is the probability it is in state D after 1 time step? After 2 time steps?

11.1. DISTRIBUTIONS EVOLVE OVER TIME 243

Exercise 11.1.9. Use the model in the transition diagram in Figure 10.3c. If initially the model is in state E, what is the probability it is in state F after 1 time step? After 2 time steps?

Exercise 11.1.10. Use the model in the transition diagram in Figure 10.3d. If initially the model is in state H, what is the probability it is in state G after 1 time step? After 2 time steps?

Exercise 11.1.11. An ion channel can be in either an open or closed state. If it is open, then it has probability 0.1 of closing in 1 microsecond; if closed, it has probability 0.3 of opening in 1 microsecond. Suppose that initially 50% of ion channels are open and 50% are closed. What fraction is open after 1 microsecond? After 2 microseconds?

Exercise 11.1.12. An individual can be either susceptible or infected; the probability of infection for a susceptible person is 0.05 per day, and the probability an infected person becoming susceptible is 0.12 per day. Suppose that initially the population is 90% susceptible and 10% infected. What fraction is susceptible after 1 day? After 2 days?

Exercise 11.1.13. The genotype of an organism can be either normal (wild type) or mutant. Each generation, a wild type individual has probability 0.03 of having a mutant offspring, and a mutant has probability 0.005 of having a wild type offspring. Suppose that initially 0.9 of the population is wild type and 0.1 is mutant. What fraction of the population is wild type after 1 generation? After 2 generations?

Exercise 11.1.14. The nAChR ion channel can be in one of three states: resting (R); closed with ACh bound (C); and open (O) with transition probabilities (per microsecond) of 0.04 (from R to C), 0.07 (from C to R), 0.12 (from C to O), and 0.02 (from O to C); the other transition probabilities are 0. Suppose that initially 3/4 of ion channels are in state R and 1/4 are in C. What fraction of the ion channels is open after 1 microsecond? After 2 microseconds?

Exercise 11.1.15. There are three kinds of vegetation in an ecosystem: grass, shrubs, and trees. Every year, 25% of grassland plots are converted to shrubs, 20% of shrub plots are converted to trees, 8% of trees are converted to shrubs, and 1% of trees are converted to grass; the other transition probabilities are 0. Suppose that initially the ecosystem is evenly split: 1/3 grass, 1/3 shrubs, and 1/3 trees. What fraction of ecosystem is covered in shrubs after 1 year? After 2 years?

11.2 Matrix multiplication in R

The easiest way to perform the cumbersome calculations for matrix multiplication is to outsource them to a computer. R provides a special operation symbol just for this purpose, which is an asterisk surrounded by percent signs: %*%. To illustrate, we will multiply the matrix A and the vector b:

$$A = \begin{pmatrix} 3 & 1 \\ -5 & 0 \end{pmatrix}; b = \begin{pmatrix} 10 \\ -2 \end{pmatrix}$$

To perform this operation in R, we must first define the matrix and the vector, and then perform multiplication.

```
A <- matrix(c(3, -5, 1, 0), nrow = 2)
print(A)
```

```
##      [,1] [,2]
## [1,]    3    1
## [2,]   -5    0
```

```
b <- c(10, -2)
c <- A %*% b
print(c)
```

```
##      [,1]
## [1,]   28
## [2,]  -50
```

As discussed in section 11.1, the probability distribution vector for a Markov model advances one time step at a time by multiplication with the transition matrix of the model. For example, let us

11.2. MATRIX MULTIPLICATION IN R

take the same transition matrix as in section 11.1, and multiply it by the initial vector prob0 = (0.5, 0.5) (which means that initially the model is in states 1 and 2 with equal probability 0.5).

```
M <- matrix(c(0.95, 0.05, 0.1, 0.9), nrow = 2)
print(M)
```

```
##      [,1] [,2]
## [1,] 0.95  0.1
## [2,] 0.05  0.9
```

```
prob0 <- c(0.5, 0.5)
prob1 <- M %*% prob0
print(prob1)
```

```
##       [,1]
## [1,] 0.525
## [2,] 0.475
```

Remember, when defining the transition matrix the element $M[i, j]$ (ith row and jth column) contains the probability of transition from state number j to state number i. Also, take care to enter the transition probabilities in the correct order, as the matrix() function by default places the element by column (first fills the first column, then the second), as you can see in the script above. The result shows that after 1 time step, the probability distribution vector changes from (0.5, 0.5) to (0.525, 0.475). What about taking many time steps?

As we know, computers can perform repetitive operations much better than humans can, so we will take advantage of their arithmetic proficiency. Since each time step involves multiplication of the current probability vector by the transition matrix, this can be done automatically with a for loop. The only difficulty is that, while the transition matrix remains the same, the probability vector needs to be updated. There are two ways of handing this: (1) replace the old vector with the new, with the disadvantage that the previous vectors all get overwritten in memory; (2) save all probability vectors in a rectangular matrix, which means we can plot all the probability vectors over time and see their evolution. The following script takes the first approach:

```
M <- matrix(c(0.95, 0.05, 0.1, 0.9), nrow = 2)
print(M)
```

```
##      [,1] [,2]
## [1,] 0.95  0.1
## [2,] 0.05  0.9
```

```
prob <- c(0.1, 0.9)
numstep <- 20
for (i in 1:numstep) {
    prob <- M %*% prob
}
print(prob)
```

```
##           [,1]
## [1,] 0.6447029
## [2,] 0.3552971
```

The script produces only the probability vector after a given number of time steps. If we want to demonstrate the evolution of the probability vectors, we need to define a rectangular matrix large enough to hold all probability vectors, starting with the initial distribution. The variable that holds this information is a matrix with as many rows as there are states (in our example, 2) and the number of columns is one more than the number of time steps. The next script accomplishes this goal:

```
M <- matrix(c(0.95, 0.05, 0.1, 0.9), nrow = 2)
print(M)
```

```
##      [,1] [,2]
## [1,] 0.95  0.1
## [2,] 0.05  0.9
```

11.3. MUTATIONS IN EVOLUTION

```
numstep <- 20
prob.vec <- matrix(0, nrow = 2, ncol = numstep + 1)
prob.vec[, 1] <- c(0.1, 0.9)
for (i in 1:numstep) {
    prob.vec[, i + 1] <- M %*% prob.vec[, i]
}
```

The key step is the command `prob.vec[ , i + 1] <- M %*% prob.vec[ , i]` which tells R to take the column number i from the matrix `prob.vec`, multiply it by the matrix `M`, and put the result into column number $i + 1$. The result is a matrix in which each column is a probability vector at a particular time. The resulting collection of vectors can be plotted over time using the function `barplot()`, as shown in Figure 11.3a. If you prefer to plot the probability vectors only at select times, for example, every other time step out of the 21 times, it is easy to do: just put in the vector of numbers you want into the time index, as shown in the code for generating Figure 11.3b. To label the times on the x-axis, use the option `names.arg`, as you can see in the code.

11.3 Mutations in an evolutionary lineage

Think of a genetic sequence (of either nucleotides or amino acids) evolving over many generations. Once in a while, a mutation will change one of the letters in the sequence; the most common mutations, as discussed in Chapter 3, are substitutions. Although each position can contain multiple letters (4 for DNA, 20 for amino acids), let us simplify the question as follows: if we know the ancestral sequence, what fraction of the sequence is unchanged after a given number of generations? To answer this question, we only need two states to describe each position in the sequence: ancestral (A) and mutant (M). The transition probability from A to M is the substitution mutation rate, which has units of mutations per nucleotide per generation. The transition probability from M to A is the rate of reversion to the ancestral state, which is reasonably assumed to be less than the overall mutation rate, since for DNA there are three options for mutation from an ancestral letter, but only one option

```
barplot(prob.vec, xlab = "time", names.arg = 0:numstep,
    ylab = "probability", cex = 1.5, cex.axis = 1.5,
    cex.lab = 1.5)
time.index <- seq(1, 21, 2)
barplot(prob.vec[, time.index], xlab = "time",
    names.arg = time.index - 1, ylab = "probability",
    cex = 1.5, cex.axis = 1.5, cex.lab = 1.5)
```

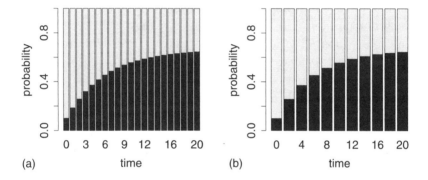

Figure 11.3: Evolution of probability vectors of a two-state Markov model over 20 time steps. (a) Vectors plotted every time step; (b) vectors plotted every other time step. (Script to generate figures is shown above the plots.)

for reversion (only one ancestral letter). Thus, under the simple assumption that all substitution mutations are equally probable, we can postulate that for a mutation rate a, the reversion rate is $a/3$.

Figure 11.4 shows the evolution of probability vectors for this model for two different values of mutation rate a. In both calculations initially 100% of the sequence is made up of ancestral letters, and then some fraction acquires mutations. Not surprisingly, if the mutation rate is greater, mutations are acquired faster, and the fraction of ancestral letter declines faster: in the plot on the left ($a = 0.0001$) more than 90% of the sequence is indistinguishable from the ancestor after 1000 generations, and in the plot on the right ($a = 0.001$) less than half remains in the ancestral state after the same time passed. This model is a simplification of the famous Jukes-Cantor model of molecular evolution, which we will investigate in section 13.4, where we will use it to predict the time of divergence of two sequences from a common ancestor.

Discussion questions:

These questions refer to the two-state mutation model described above.

Discussion 11.3.1. What does the Markov property mean for this model? How realistic do you think it is for real genetic sequences?

Discussion 11.3.2. What does this model assume about the relationship between different positions in the sequence? Is this a realistic assumption?

Discussion 11.3.3. The probability vectors plotted in Figure 11.4 are deterministic; that is, they can be predicted exactly from an initial probability vector. Does this mean that any sequence (which obeys the assumptions of the model) will match the predicted ancestral fraction? Which sequence do you expect to better match the prediction: a short one or a long one?

Discussion 11.3.4. What do you expect would happen if the calculation continued for many more generations? In other words, for a DNA sequence that is very far removed from its ancestor, what fraction of letters do you expect will match?

11.4 Computational projects

In the following projects you will generate a sequence of probability vectors and store it in a rectangular matrix (e.g., `prob.vec`). Starting with a given initial provability, you can multiply it by the transition matrix to calculate the next probability vector. To do this repeatedly for a set number of time steps it is best to use a for loop, as shown in the script in section 11.3.

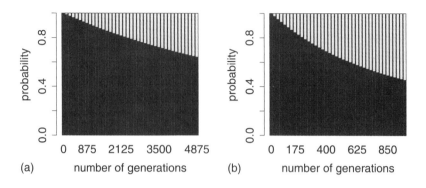

Figure 11.4: The fraction of the sequence identical to the ancestral sequence after a certain number of generations for two different mutation rates: (a) $a = 0.0001$ for 5000 generations; (b) $a = 0.001$ for 1000 generations.

11.4.1 probability vectors of a two-state model

In this section you will calculate the probability distribution vectors of the two-state SI model with the same transition probabilities as in the previous chapter's assignment (per day): 0.04 (from S to I) and 0.12 (from I to S). In tasks 4 and 5 you will generate a sequence of probability vectors stored in a 2-by-(numsteps+1) matrix, where numsteps is the number of time steps, and plot them to observe how the probability distribution changes over time.

Tasks

1. Write down the transition matrix for the model, and assign it to matrix M in R. Remember that all the probabilities of transitioning from a state have to add up to 1.

2. Suppose that a person is initially healthy (i.e., the probability of susceptible is 1, and the probability of infected is 0). Use matrix multiplication to calculate the probability the person is infected after 1 day and after 2 days.

11.4. COMPUTATIONAL PROJECTS

3. Change the initial probability to the person being infected (probability of susceptible is 0, probability of infected is 1). Use matrix multiplication to calculate the probability the person is infected after 1 day and after 2 days.

4. Write a script to calculate the distribution of infected and susceptible individuals for any given number of days (use a for loop). Using the script, calculate the probability distribution vectors for 100 days, using both initial probability vectors from above, save them as matrices, and plot the probability vectors over time using `barplot()`. How different are the probability distributions after 100 days?

5. Change the infection rate (transition from S to I) to 0.2, repeat the computations in the previous task, and plot the two probability distributions over time using `barplot()`. Are the probability distributions based on the two initial vectors different from each other? How is their time evolution different from the distributions above?

11.4.2 probability vectors of a three-state model

In this project we return to the three-state model of nicotinic acetylcholine receptor (nAChR) described in Chapter 10. As before, set the transition probabilities to the following: 0.04 (from R to C), 0.07 (from C to R), 0.12 (from C to O), and 0.02 (from O to C); the other transition probabilities are 0. In tasks 5 and 6 you will generate a sequence of probability vectors, stored in a 3-by-(numsteps+1) matrix, where numsteps is the number of time steps, and plot them to observe how the probability distribution changes over time.

Tasks

1. Write down the transition matrix for the model, and assign it to a matrix M. Remember that all the probabilities of transitioning from a state have to add up to 1.

2. Suppose that the ion channel is initially in state R. Use matrix multiplication to calculate the probability that the channel is open after 1 and 2 microseconds.

3. Suppose that the ion channel is initially in state C. Use matrix multiplication to calculate the probability that the channel is open after 1 and 2 microseconds.

4. Suppose that the ion channel is initially in state O. Use matrix multiplication to calculate the probability that the channel is open after 1 and 2 microseconds.

5. Suppose that an ion channel is initially in state R; use a script to propagate the probability vector for 100 microseconds, and plot the probability vector over time using `barplot()`. Repeat the calculations starting with the channel in state C and again in state O. How different are the probability distributions based on the three initial vectors after 100 time steps? Does the probability distribution converge to an equilibrium and if so, over how many time steps?

6. Change the opening rate (transition from C to O) to 0.02, repeat the computations in the previous task, and plot the three probability distributions over time using `barplot()`. How different are the probability distributions based on the three initial vectors after 100 time steps? How is their time evolution different from the distributions in task 5?

Chapter 12

Stationary distributions of Markov chains

> *The tears of the world are a constant quantity. For each one who begins to weep somewhere else another stops. The same is true of the laugh.*
> —Samuel Beckett, *Waiting for Godot*

In Chapter 11 you learned to compute the distributions of Markov models, bringing a measure of predictability to the randomness. Using matrix multiplication repeatedly we can compute the distribution for any given time, and observe how probability vectors evolve. You may have noticed that in the examples in the computational projects, the probability vectors tended to approach some particular distribution and then essentially remain the same. It turns out that Markov chains have special stationary distributions at which the transitions are perfectly balanced, so the probabilities of each state remain the same. In this chapter you will study the stationary distributions of Markov chains and learn to do the following:

1. calculate the stationary distribution of small Markov models on paper,

2. tell whether a Markov chain converges to a stationary distribution,

3. run multiple simulations in R and observe convergence to a stationary distribution, and

4. understand the concepts of Hidden Markov Models.

12.1 The origins of Markov chains: A feud and a poem

The idea of chains of random variables that depend on each other was born of a feud. In the late nineteenth century probability theory had made great strides, both in theory and in groundbreaking applications to physics, such as the work of Boltzmann in thermodynamics. Randomness and its mysteries had become a fashionable topic of conversation outside the confines of mathematics classrooms and conferences. Sociological studies were published that claimed to show that the behavior of a large number of people was predictable due to the Law of Large Numbers. The mathematician and self-styled philosopher P. A. Nekrasov published a paper in 1902 that made an audacious leap of logic: he claimed that since human beings were subject to the Law of Large Numbers, and that law requires independence between constituent random variables, humans must have been endowed with free will, in agreement with his devout Russian Orthodox beliefs. The argument is questionable both mathematically and theologically, and it especially grated on another mathematician, A. A. Markov (Hayes 2012).

Markov was a great mathematician as well as a malcontent. In contrast with Nekrasov, he was neither a monarchist nor a devout Orthodox believer, and even asked to be excommunicated from the church after it expelled the great writer Tolstoy for heresy. Markov already disdained Nekrasov both personally and professionally, and the paper inspired him to action. After several years of work, he published a paper titled "An extension of the Law of Large Numbers to Quantities Which Depend on Each Other" (Markov 1913), which founded the concept of Markov chains. As the title states, it provided a counterexample to Nekrasov's claim that the predictability of the behavior of large number of random variables implied their

12.1. ORIGINS OF MARKOV CHAINS

independence. Markov showed that variables that depend on each other can also behave in a predictable manner in large numbers.

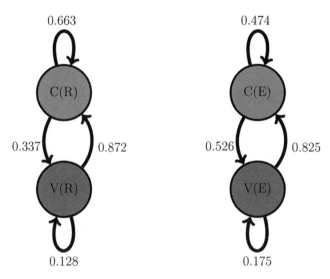

Мой дядя самых честных правил,
Когда не в шутку занемог,
Он уважать себя заставил
И лучше выдумать не мог.

My uncle, man of firm convictions...
By falling gravely ill, he's won
A due respect for his afflictions
The only clever thing he's done.

Figure 12.1: Two Markov models based on the text of *Eugene Onegin*, with states denoting consonants (C) and vowels (V); model based on the Russian text is on the left and the one based on English on the right (Hayes 2012). The first four lines of the poem in the original (Pushkin 1832) and in English translation (Pushkin 2009) are below their respective diagrams.

In addition to inventing the mathematical concepts, Markov was the first to use his chains of random variables to make a Markov model. In his 1913 paper (Markov 1906) he proposed a model based on the classic Russian poem *Eugene Onegin* by Pushkin. To make the task manageable he divided the letters into two categories: consonants and vowels, discarding spaces, punctuation, and two Russian letters that make no sound. To calculate the transition probabilities between the two states, Markov took the first 20,000 letters of the poem and counted by hand the fraction of vowels that were followed by vowels, and the fraction of consonants that were followed by consonants. He then built the first two-state text-based model, foreshadowing models of bioinformatics used now to analyze genome structure.

Figure 12.1 shows two models based on 20,000 letters of *Eugene Onegin* in Russian and in English translation. The resulting transition probabilities are different than those computed by Markov in his paper: whereas in the English text the probability of a vowel following another vowel is 0.175, in the original Russian it is 0.128; in English the probability of a consonant following another consonant is 0.474, while in Russian it is 0.663. Clearly, Russian words contain more consonant clusters and fewer vowels next to each other. In both cases, the state of the previous letter affects the probability of the next letter being a vowel. Remarkably, the distribution of consonants and vowels is predictable in any sufficiently long piece of text: in English it is about 39% vowels and 61% consonants, and in Russian it is about 28% vowels and 72% consonants. This is an example of the main result of the first Markov chain paper: large numbers of interconnected random variables converge to a predictable distribution, called the *stationary distribution*.

12.2 Stationary distributions

12.2.1 definition of stationary distribution

What happens when we extend our calculation of the probability distribution vectors of a Markov chain over a long time? Let us consider the cell cycle model with states Q and R (see Chapter 11). We have seen the state sequences of a single cell over time, so consider what happens to a population of cells. The basic question is: given that all cells start out in a particular state (e.g., R), what fraction of cells is in state R after a certain number of time steps? Figure 12.2 shows the result of propagating the QR model for 30 time steps, starting with two different initial distributions. You can try this at home yourself, starting with different initial distributions, and see that all of them converge over time to the same fraction of Q and R. This is called the *stationary distribution* of the Markov chain.

12.2. STATIONARY DISTRIBUTIONS

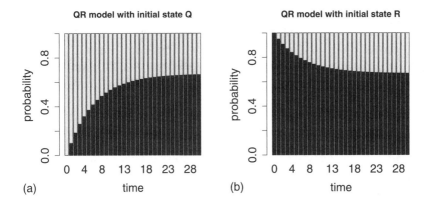

Figure 12.2: Probability distributions converge to the same distribution starting from two different initial distributions: (a) $P(0) = (0, 1)$; (b) $P(0) = (1, 0)$.

Definition 12.1. For a finite-state Markov model with transition matrix M a *stationary or equilibrium distribution* is a vector $\vec{P}_s$ that has all non-negative elements which add up to 1, and satisfies

$$\vec{P}_s = M \times \vec{P}_s$$

The definition says that a probability vector that is unchanged by multiplication by the transition matrix will remain stationary over time (Feller 1968).

Example. The stationary distribution vector can be calculated analytically from definition 12.1. Let us find the stationary vector $\vec{P}_s$ for the QR cell model with components P_Q and P_R (the fractions of quiescent and replicating cells in the stationary distribution):

$$\begin{pmatrix} P_Q \\ P_R \end{pmatrix} = \begin{pmatrix} 0.95 & 0.1 \\ 0.05 & 0.9 \end{pmatrix} \begin{pmatrix} P_Q \\ P_R \end{pmatrix} = \begin{pmatrix} 0.95 P_Q + 0.1 P_R \\ 0.05 P_Q + 0.9 P_R \end{pmatrix}$$

This means there are two equations to solve for two variables. It turns out that they are equivalent:

$$0.95 P_Q + 0.1 P_R = P_Q \Rightarrow 0.1 P_R = 0.05 P_Q \Rightarrow P_R = 0.5 P_Q$$

$$0.05P_Q + 0.9P_R = P_R \Rightarrow 0.05P_Q = 0.1P_R \Rightarrow 0.5P_Q = P_R$$

Both equations say that in the stationary distribution, there are twice as many quiescent cells as replicating ones. If we add the condition that $P_Q + P_R = 1$, then we have the exact solution:

$$\vec{P}_s = \begin{pmatrix} P_Q \\ P_R \end{pmatrix} = \begin{pmatrix} \frac{2}{3} \\ \frac{1}{3} \end{pmatrix}$$

Thus in a large population of cells in the cell cycle model, a population with 2/3 quiescent and 1/3 replicating cells is stationary. This does not mean that each individual cell remains in the same state! Each cell still randomly transitions between the two states, but the number of cells switching to the quiescent state is balanced by the number of cell switching out of the state, so the net distribution remains the same. We will observe this using simulations with multiple individual cells in section 12.3.

Math exercises:

For the following Markov models, (a) write down the transition matrix M; (b) find the stationary probability distribution on paper; (c) use matrix multiplication in R to check that the distribution you found satisfies the definition of stationary distribution.

Exercise 12.2.1. Use the model in the transition diagram in Figure 10.3a.

Exercise 12.2.2. Use the model in the transition diagram in Figure 10.3b.

Exercise 12.2.3. Use the model in the transition diagram in Figure 10.3c.

Exercise 12.2.4. Use the model in the transition diagram in Figure 10.3d.

Exercise 12.2.5. An ion channel can be in either an open or closed state. If it is open, then it has probability 0.1 of closing in 1 microsecond; if closed, it has probability 0.3 of opening in 1 microsecond.

12.2. STATIONARY DISTRIBUTIONS

Exercise 12.2.6. An individual can be either susceptible or infected; the probability of infection for a susceptible person is 0.05 per day, and the probability an infected person becoming susceptible is 0.12 per day.

Exercise 12.2.7. The genotype of an organism can be either normal (wild type) or mutant. Each generation a wild type individual has probability 0.03 of having a mutant offspring, and a mutant has probability 0.005 of having a wild type offspring.

Exercise 12.2.8. A gene is is either expressed (On) or is not expressed (Off) by a stochastic mechanism. In the On state, it has probability 0.3 per minute of turning off, and in the Off state, it has probability 0.02 per minute of turning on.

Exercise 12.2.9. The nAChR ion channel can be in one of three states: resting (R), closed with ACh bound (C), and open (O) with transition probabilities (per microsecond) of 0.04 (from R to C), 0.07 (from C to R), 0.12 (from C to O) and 0.02 (from O to C); the other transition probabilities are 0.

Exercise 12.2.10. There are three kinds of vegetation in an ecosystem: grass, shrubs, and trees. Every year 25% of grassland plots are converted to shrubs, 20% of shrub plots are converted to trees, 8% of trees are converted to shrubs, and 1% of trees are converted to grass; the other transition probabilities are 0.

12.2.2 condition for unique stationary distribution

You now know how to find stationary distributions from the definition, and we saw in the computer calculations how probability vectors converge to these distributions over time. However, just because there is a stationary distribution does not mean that the model will converge to it. Here is a very simple example: a two-state model where transitions between the two states both have probability 1. Here is its transition matrix:

$$M = \begin{pmatrix} 0 & 1 \\ 1 & 0 \end{pmatrix}$$

It is clear that the vector $P_s = (0.5, 0.5)$ is a stationary distribution. However, if we start with an initial distribution that is different from P_s, the distribution does not converge to any stationary vector. For instance, if the initial distribution is all in state 1, the distribution vector will keep flipping between $(1, 0)$ and $(0, 1)$ forever.

All Markov models have at least one stationary distribution, as we will see in Chapter 13, but there could be more than one. Here is another simple example with two states where the transition probabilities are exactly zero:

$$M = \begin{pmatrix} 1 & 0 \\ 0 & 1 \end{pmatrix}$$

You can check for yourself that both vectors $(1, 0)$ and $(0, 1)$ satisfy the definition of stationary distribution. In fact, any vector is stationary, since nothing can leave either state, so there are infinitely many stationary distributions of this very boring model.

So can we avoid the weirdness of either cycling around endlessly, as in the first example, or having multiple stationary distributions? We first need to define a new concept of *communication* between states to express the condition for a unique stationary distribution.

Definition 12.2. Two states i and j *communicate* with each other if there is nonzero probability of transition from i to j after some number of time steps, as well as for the probability of transition from j to i.

Determining whether two states communicate with each other can be done by checking the flow diagram of the model. As long as there is a path from state i to state j using nonzero probability transitions, and vice versa, the two states communicate. States that all communicate with one another form a *class*, which is a mathematical term that I think is intuitive enough not to bother with formalities.

Definition 12.3. A Markov chain is *irreducible* if all its states belong to one class, that is, if all states communicate.

Markov states can be divided into two types: recurrent and transient.

12.2. STATIONARY DISTRIBUTIONS

Definition 12.4. A state i in a Markov chain is called *recurrent* if the random variable has probability 1 of returning to state i at some future time. Otherwise it is called *transient*.

Example. A Markov chain with two states where X transitions from state 1 to 2 with probability 1, and stays in state 2 with probability 1, has the following transition matrix:

$$M = \begin{pmatrix} 0 & 0 \\ 1 & 1 \end{pmatrix}$$

In this model state 1 is transient, because there is no probability of returning to it, while state 2 is recurrent—in fact, it is a special kind of recurrent state called *absorbing*.

Finally, there is one more notion necessary to describe Markov states: periodicity.

Definition 12.5. A state i in a Markov chain is *periodic with period* n if after some time k, the probability of return to the state is nonzero only for integer multiples of n. A state with periodicity 1 is called *aperiodic*.

Example. The same Markov chain introduced above is an example of periodicity, because the probability of return to either state is nonzero only for multiples of 2:

$$M = \begin{pmatrix} 0 & 1 \\ 1 & 0 \end{pmatrix}$$

Notice that both of these states are recurrent as well as periodic, but they are not absorbing.

Finally, we have the tools to state the conditions that guarantee that a Markov chain has only one stationary distribution and that all initial distributions converge to it (Feller 1968).

Theorem 12.1. A Markov chain in which all states communicate with one another (irreducible) and all states are recurrent and aperiodic has a unique stationary distribution, and all initial distributions converge to it.

The three different Markov chains in the above examples do not satisfy the requirements of this theorem. The first one,

$$M = \begin{pmatrix} 0 & 1 \\ 1 & 0 \end{pmatrix}$$

has periodic states, whereas the theorem requires aperiodicity. The second one,

$$M = \begin{pmatrix} 1 & 0 \\ 0 & 1 \end{pmatrix}$$

has aperiodic states, but they do not communicate (the Markov chain is not irreducible). The third one,

$$M = \begin{pmatrix} 0 & 0 \\ 1 & 1 \end{pmatrix}$$

has two problems: the states don't communicate, and state 1 is not recurrent.

However, the QR model that we have been using for the past three chapters (see section 10.1), with transition matrix

$$M = \begin{pmatrix} 0.95 & 0.1 \\ 0.05 & 0.9 \end{pmatrix}$$

fits the bill, as it is irreducible, recurrent, and aperiodic, and as promised, it has a unique stationary distribution.

Math exercises:

The transition diagrams shown in Figure 12.3 show arrows only for transitions that are possible; all other transitions have probability zero. For each model do the following: (a) classify each state as transient or recurrent; (b) decide whether the states are periodic or aperiodic; (c) determine whether the Markov chain is ergodic and irreducible; (d) decide whether it has a unique stationary distribution.

Exercise 12.2.11. Use the transition diagram in Figure 12.3a.

Exercise 12.2.12. Use the transition diagram in Figure 12.3b.

Exercise 12.2.13. Use the transition diagram in Figure 12.3c.

Exercise 12.2.14. Use the transition diagram in Figure 12.3d.

12.2. STATIONARY DISTRIBUTIONS

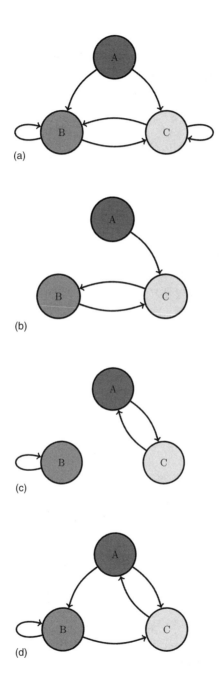

Figure 12.3: Transition diagrams for three-state Markov models; arrows indicate transitions with nonzero probabilities.

12.3 Multiple random simulations in R

In this section we perform simulations for multiple manifestations of a Markov model and observe how the proportions of different states evolve over time. We simulate the QR cell cycle model from Chapter 9, run the simulation code for simulating a single cell, and repeat it multiple times. We assume that different cells do not influence one another, so each simulation can be repeated without reference to the others. The best way to repeat the same operations multiple times is to use a for loop, so in the script below, I put another for loop around the script for simulating a single cell, with the new indexing variable going through different cells. Within the outer loop there is the internal loop with the indexing variable going through time steps. The generated states (1 for Q and 2 for R) are stored in a matrix with the number of columns given by the number of cells, and the number of rows by the number of time steps plus one (to leave room for the initial state).

```
prob1 <- 0.05
# set transition probability from state 1 to 2
prob2 <- 0.1
# set transition probability from state 2 to 1
nsteps <- 40   # set number of time steps
ncells <- 30   # set number of cells
states <- matrix(1, nrow = ncells, ncol = nsteps + 1)
    # initialize states matrix
for (j in 1:ncells) {
    # go through all the cells go through time steps
    for (i in 1:nsteps) {
        decider <- runif(1)   # generate a random number
        if (states[j, i] == 1) {
            if (decider < prob1) {
                # transition from 1 to 2
                states[j, i + 1] <- 2   # new state is 2
            } else {
                states[j, i + 1] <- 1
                # new state is still 1
```

12.4. BIOINFORMATICS AND HMMS

```
        }
    } else {
        if (decider < prob2) {
            # transition from 2 to 1
            states[j, i + 1] <- 1   # new state is 1
        } else {
            states[j, i + 1] <- 2
            # new state is still 2
        }
    }
  }
}
```

The number of cells in states 1 and 2 at each time step is tabulated and then plotted using `barplot()` in a separate script that isn't shown. The results are shown in Figure 12.4 for two simulations with a total of 30 individual cells, starting with all cells in state 1 (Figure 12.4a) and starting with all states in state 2 (Figure 12.4b). Although in both plots the numbers of cell in Q tend to approach approximately 20 out of 30, the numbers bounce around a lot due to the inherent randomness of the simulation. One can make the frequencies less volatile by increasing the cell population. Figure 12.5 shows the results of simulations with 200 cells, which show the numbers of cells in state 1 approach 2/3 of 200 (around 134), with considerably less variance. Finally, the counts for simulations with 1000 cells are shown in Figure 12.6, and the counts in that figure are much closer to what may be called convergence to the stationary distribution, which you can see by comparing them with the plots of the probability distribution vectors in Figure 12.2 in section 12.2. This illustrates what Markov called the Law of Large Numbers for a chain of variables that depend on one another.

12.4 Bioinformatics and Markov models

In section 12.1 we saw a simple Markov model for a string of characters, which was used to model a poetic text in Russian. While it did provide some information about the distribution of the vowels

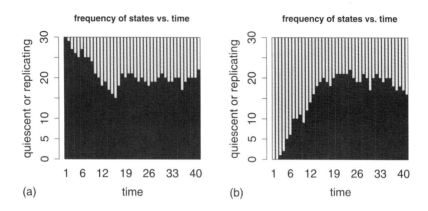

Figure 12.4: Counts of simulated cells in states Q and R (states 1 and 2, respectively) out of a total count of 30 converge over time to the same distribution starting from (a) all in state Q; (b) all in state R.

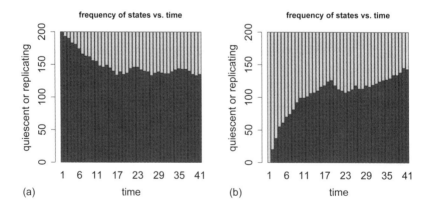

Figure 12.5: Counts of simulated cells in states Q and R (states 1 and 2, respectively) out of a total count of 200 converge over time to the same distribution starting from (a) all in state Q; (b) all in state R.

12.4. BIOINFORMATICS AND HMMS

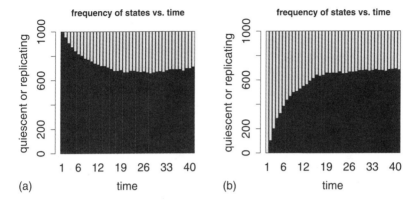

Figure 12.6: Counts of simulated cells in states Q and R (states 1 and 2, respectively) out of a total count of 1000 converge over time to the same distribution starting from (a) all in state Q; (b) all in state R.

and consonants in the text—for instance, that it is substantially more likely that a vowel is followed by a consonant than by another vowel—the usefulness of the model is limited. However, analysis of strings of characters is a crucial component of modern biology, which is awash in sequence data: DNA, RNA, and protein sequences from different organisms are pouring into various data bases. Markov models have become indispensable for making sense of sequence data.

One of the major problems in *bioinformatics* is identifying portions of the genome that code for proteins (Pevsner 2009). A genomic sequence consists of four letters, but their meaning and function depends on where they are and how they are used. Some parts of the genome (in humans, over 90%) are not part of a gene, and the DNA sequence is never translated into an amino acid sequence. Others are genes, which are continuous chunks of DNA sequence that are flanked by a promotor sequence and regulatory region (which controls when a gene is *expressed*), followed by the gene proper, which is transcribed into RNA and then translated into amino acids. Some parts used to be genes but are no longer in use; these are called *pseudogenes*. They can be difficult to distinguish from functional genes, because their sequences still have similar features, including the proximity of promoters, and regulatory and coding regions.

Within the borders of a gene are other divisions. In eukaryotic genomes, after the gene sequence is transcribed into RNA, some portions called *introns* are cut out; then the remaining pieces called *exons* are spliced together and only then translated into protein sequences. The role of introns in biology is a topic of ongoing research, since it seems rather wasteful to transcribe portions of genes, which are sometimes considerably longer than the protein-coding exons, only to discard them later. The problem of identifying introns in a gene is important.

Markov models are used to determine the structure behind the sequence of letters. Based on known sequences of exons and introns, one can generate a *Hidden Markov Model* (HMM) that connects the DNA sequence with its underlying meaning: whether it is part of an exon or an intron. These models are more complex than the plain Markov models that we have studied; they involve two sets of states: the hidden ones, like introns and exons, which are not observable, and the observations, such as the four nucleotides (A,T,G,C). There are also two sets of transition probabilities: the transition probabilities between hidden states, and the *emission probabilities*, which are the probabilities that a hidden state produces a particular observation.

Figure 12.7 shows an example of such a model for gene structure. The HMM has three hidden states: E (exon), 5 (the 5' boundary of an intron), and I (intron). Each of these states has its own probability distribution of nucleotides (letters in the sequence), with Exons containing equal proportions of all four letters, the 5' almost always being a G, and the Introns containing four times as many As and Ts as Gs and Cs. Each hidden state has its own probability of "emitting" a letter, so one can devise algorithms for finding the most probable string of hidden states based on an observed sequence of nucleotides. HMM enables intron-hunting to be done in a systematic manner, although, as with any random model, the results are never certain.

12.4. BIOINFORMATICS AND HMMS

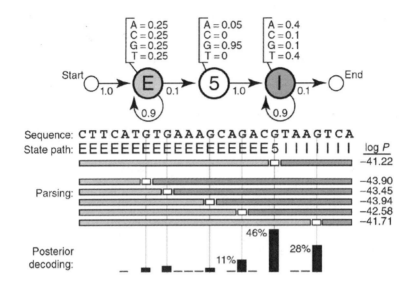

Figure 12.7: Diagram of a simple Hidden Markov Model for a eukaryotic gene. Figure from Eddy (2004), used by permission.

Discussion questions:

The following questions refer to the paper "What Is a Hidden Markov Model?" (Eddy, 2004).

Discussion 12.4.1. What does the Markov property mean for HMMs presented in this paper? How reasonable is it for an actual genetic sequence?

Discussion 12.4.2. HMMs can predict the "best state path" or the sequence of hidden states with the highest probability. Why is a single state path often not sufficient to answer questions of interest?

Discussion 12.4.3. What additional assumptions does the HMM in Figure 12.7 make about the distribution of letters in an exon or intron? Comment on the biological implications.

Discussion 12.4.4. What bioinformatics problems are HMMs best suited for? What are some of their drawbacks?

12.5 Computational projects

In the projects below you will generate multiple simulations over a number of time steps. The states of n simulations for m days should be stored in a matrix with n rows and $m + 1$ columns, because we begin with the initial state on day 1 and then add m more days. Suppose that this matrix is called states, and we are simulating the two-state disease models; then the the state (1 or 2) of person number 3 after 10 days will be recorded in states[3,11]. To plot the histogram of the states at a particular time, one can use the table() function together with barplot(), for instance, to plot the distribution on day 10:

```
barplot(table(states[, 10]),
        main = "distribution at time 10")
```

To plot the distribution of states over all different times, using table() on the entire matrix doesn't work. Instead, one has to generate a vector of counts for each time step and then plot it using barplot(). The script below shows how to do this, given a matrix of states (states), the number of time steps (numsteps), and the number of states (numstates), which is 2 for the SI model.

```
state.count <- matrix(0, nrow = numstates,
    ncol = nsteps + 1)
for (k in 1:(numsteps + 1)) {
    state.count[, k] <- tabulate(states[, k],
    nbins = numstates)
    # count the states at timestep
}
barplot(state.count, names.arg = 1:(numsteps + 1),
    ylab = "S or I")
```

12.5.1 multiple two-state model simulations

In this section you will generate multiple simulations of the two-state SI model with the same transition probabilities as in Chapter 11's is assignment: 0.04 (from S to I) and 0.12 (from I to S). Take

12.5. COMPUTATIONAL PROJECTS

your code for the simulation of this model from Chapter 10, and add another for loop around your script to generate a two-dimensional matrix of states.

Tasks

1. Simulate 100 individuals for 20 days (generating 100 separate state strings of length 21). Simulate the state histories of 100 individuals for 20 days (generating 100 separate state strings), starting with all 100 individuals in state S (number 1). Plot the histograms (frequencies of states 1 and 2) over time using the code provided above. Describe the behavior of the histograms over time. Does the distribution after 20 days remain the same if you run your script multiple times?

2. Repeat the previous task, but set the initial states of all 100 individuals to I (state number 2). Again, plot the histograms (frequencies of states 1 and 2), and describe the behavior of the histogram over time. Does the distribution between 1 and 2 look the same as in the previous task after 20 days?

3. Change the infection rate (transition from S to I) to 0.2, and simulate a population of 100 individuals over 20 days, starting with everyone in the susceptible state, and plot the histograms over time. Describe the behavior of the histogram over time. Does the distribution after 20 days remain the same if you run your script multiple times? Do the histograms converge to the same distribution as in task 1?

4. Keep the same transition probability as in task 3, and simulate a population of 100 individuals over 20 days, starting with everyone in the infected state, and plot the histograms over time. Describe the behavior of the histogram over time. Does the distribution after 20 days remain the same if you run your script multiple times? Do the histograms converge to the same distribution as in task 2?

12.5.2 multiple three-state model simulations

In this project we return to the three-state model of nicotinic acetylcholine receptor (nAChR) we saw in the previous chapters. As before, set the transition probabilities to the following: 0.04 (from R to C), 0.07 (from C to R), 0.12 (from C to O), and 0.02 (from O to C); the other transition probabilities are 0. Take your code for simulation of this model model from Chapter 10, and add another for loop around your script to generate a two-dimensional matrix of states.

Tasks

1. Simulate 300 ion channels for 100 microseconds, with all 300 channels initially in state C, and plot the histograms over time using the sample code above. Describe the behavior of the histogram over time. Does the distribution after 100 microseconds remain the same if you run your script multiple times?

2. Repeat the last task, but set the initial states of all ion channels to C. Again, plot the histograms and describe the behavior of the histogram over time. Does the distribution look the same as in the previous task after 100 microseconds? Has the distribution converged?

3. Repeat the last task, but set the initial states of all ion channels to O. Again, plot the histograms, and describe the behavior of the histogram over time. Does the distribution look the same as in the previous task after 100 microseconds? Has the distribution converged?

4. Change the opening rate (transition from C to O) to 0.02, and repeat the computations in task 1 with initial states set to C. Plot the the resulting histogram over time, and describe the behavior of the histogram over time. Does the distribution after 100 microseconds remain the same if you run your script multiple times? Compare the distribution you observe with the distribution in task 1.

12.5. COMPUTATIONAL PROJECTS

5. Repeat the previous task, but set the initial states of all ion channels to C. Again, plot the histograms, and describe the behavior of the histogram over time. Does the distribution look the same as in the previous task after 100 microseconds? Has the distribution converged?

6. Repeat the last task, but set the initial states of all ion channels to O. Again, plot the histograms, and describe the behavior of the histogram over time. Does the distribution look the same as in the previous task after 100 microseconds? Has the distribution converged?

Chapter 13

Dynamics of Markov models

No hour is ever eternity, but it has its right to weep.
—Zora Neale Hurston, *Their Eyes Were Watching God*

You have learned several approaches for analyzing the behavior of Markov models. We know that many Markov models converge to a single stationary distribution over time. For many biological questions, however, the stationary distribution itself is not very interesting, but what matters is how fast the probability distribution converges. In this chapter you will encounter more advanced tools for analyzing matrices, which will enable us to answer that question. This approach will be illustrated by application to the determination of evolutionary distance based on sequence data. In this chapter you will learn to do the following:

1. understand the meaning of eigenvalues and eigenvectors,

2. calculate eigenvalues and eigenvectors of 2-by-2 matrices,

3. calculate the mixing times of Markov chains, and

4. calculate the phylogenetic distance between two DNA sequences.

13.1 Phylogenetic trees

Over many generations, genomes of living creatures accumulate random mutations, as discussed in Chapters 3 and 6. Each individual genome in a population has its own polymorphisms, and thus some are more advantageous for survival in a particular environment than others. The process of *natural selection* is stochastic, and the notion of "survival of the fittest" is not a guarantee that the best-adapted always outcompete the rest—sometimes a more fit individual has a bad day and can't find any food or gets eaten by a predator. However, over time the alleles that are advantageous have a better chance of survival and become more common in the population, while other alleles become rare or vanish (Futuyma 2009). This process never stops, because the environmental conditions change, and new mutations arise, so at any point in time the individual genomes in a species have some variations, although the nearly all the sequence of their genomes are identical.

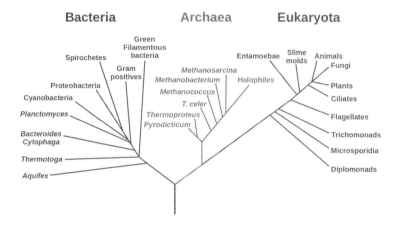

Figure 13.1: Phylogenetic tree for life forms on Earth. Figure by Eric Gaba (NASA Astrobiology Institute) in public domain via Wikimedia Commons.

One may describe the collection of genomes in a population or a species in terms of the most common alleles; this is roughly what we call the human genome, or the elephant genome, or the rice genome. Once we have determined a consensus genome sequence

13.1. PHYLOGENETIC TREES

for a species, it can be used to pose and answer questions about its heritage. If the genomes of two species are more similar to each other than to a third, it is likely that the similar pair diverged more recently than the third one. This information allows one to build *phylogenetic trees* that visually illustrate the evolutionary history of a collection of species, with each fork in the tree representing the splitting of lineages. Interpreting phylogenetic tress is fairly straightforward: species or clades (a collection of species that make an evolutionary unit) that are closely related are directly connected to a common ancestor, while the path between those that are more distantly related passes through multiple forks before reaching the common ancestor. Some trees also incorporate time information as branch lengths, with longer branches indicating that more time has passes from a divergence event.

Figure 13.1 shows the phylogenetic tree for the life forms on Earth, divided into the three major kingdoms: Bacteria, Archaea, and Eukaryota (the latter includes all multicellular lifeforms—plants, animals, and fungi). The nodes (end points) show the major groupings of life existing today, and the branches show the order of evolutionary divergence of lineages, starting with the hypothesized root of the tree at the bottom, known as the last universal common ancestor, or LUCA (Futuyma 2009). The order of splitting and the grouping of the nodes was determined by molecular sequence data (in particular, ribosomal RNA) that has become available in the past 20 years.

In the past biologists studied observable traits of different life forms, such as anatomy, physiology, or developmental features, and determined similarity from these data. However, the wealth of molecular sequence data has offered a great amount of evidence, which is the primary material of evolution. Phylogeny, particularly that of unicellular organisms, has been revolutionized by these data; we now know that prokaryotes are divided into the kingdoms of Archaea and Bacteria, which are evolutionarily more distant than humans are from fungi. Sequence data are quantitative and they require mathematical models to interpret them and to infer phylogenies. In the last section of this chapter I introduce mathematical tools that connect related sequences to the evolutionary divergence from their common ancestor.

13.2 Eigenvalues and eigenvectors in Markov models

13.2.1 basic linear algebra terminology

In the past two chapters you have seen matrices and learned the definition of matrix multiplication, but now we are ready to go deeper into the branch of mathematics studying matrices and their generalizations, called *linear algebra*. It is fundamental to both pure and applied mathematics (Strang 2005), and its tools are used in countless applications and fields. Let us define two useful numbers that help describe the properties of a matrix.

Definition 13.1. The *trace* τ of a matrix A is the sum of the diagonal elements: $\tau = \sum_i A_{ii}$. The *determinant* Δ of a 2-by-2 matrix A is given by the following: $\Delta = ad - bc$, where

$$A = \begin{pmatrix} a & b \\ c & d \end{pmatrix}$$

For larger matrices, the determinant is defined recursively in terms of 2-by-2 submatrices of the larger matrix, but I will not give the full definition here.

In this section you will learn to characterize square matrices by finding special numbers and vectors associated with them. At the core of this analysis lies the concept of a matrix as an *operator* that transforms vectors by multiplication. To be clear, in this section we take as default that the matrices A are square, and that vectors $\vec{v}$ are column vectors, and thus will multiply the matrix on the right: $A \times \vec{v}$.

A matrix multiplied by a vector produces another vector, provided the number of columns in the matrix is the same as the number of rows in the vector. This can be interpreted as the matrix transforming the vector $\vec{v}$ into another vector: $A \times \vec{v} = \vec{u}$. The resultant vector $\vec{u}$ may or may not resemble $\vec{v}$, but there are special vectors for which the transformation is simple.

Example. Let us multiply the following matrix and vector (specially chosen to make a point):

13.2. EIGENVALUES OF MARKOV MODELS

$$\begin{pmatrix} 2 & 1 \\ 2 & 3 \end{pmatrix} \begin{pmatrix} 1 \\ -1 \end{pmatrix} = \begin{pmatrix} 2-1 \\ 2-3 \end{pmatrix} = \begin{pmatrix} 1 \\ -1 \end{pmatrix}$$

We see that this particular vector is unchanged when multiplied by this matrix, or we can say that the matrix multiplication is equivalent to multiplication by 1. Here is another such vector for the same matrix:

$$\begin{pmatrix} 2 & 1 \\ 2 & 3 \end{pmatrix} \begin{pmatrix} 1 \\ 2 \end{pmatrix} = \begin{pmatrix} 2+2 \\ 2+6 \end{pmatrix} = \begin{pmatrix} 4 \\ 8 \end{pmatrix}$$

In this case, the vector is changed, but only by multiplication by a constant (4). Thus the geometric direction of the vector remains unchanged.

Generally, a square matrix has an associated set of vectors for which multiplication by the matrix is equivalent to multiplication by a constant. This can be written down as a definition.

Definition 13.2. An *eigenvector* of a square matrix A is a vector $\vec{v}$ for which matrix multiplication by A is equivalent to multiplication by a constant. This constant λ is called the *eigenvalue* of A corresponding to the eigenvector $\vec{v}$. The relationship summarized by.

$$A \times \vec{v} = \lambda \vec{v} \tag{13.1}$$

Note that this equation combines a matrix A, a vector $\vec{v}$, and a scalar λ, and that both sides of the equation are column vectors. This definition is illustrated in Figure 13.2, showing a vector ($\vec{v}$) multiplied by a matrix A, and the resulting vector $\lambda\vec{v}$, which is in the same direction as $\vec{v}$, due to scalar multiplying all elements of a vector, thus either stretching it if $\lambda > 1$ or compressing it if $\lambda < 1$. This assumes that λ is a real number, which is not always the case, but we will leave that complication aside for the purposes of this chapter.

The definition does not specify how many such eigenvectors and eigenvalues can exist for a given matrix A. There are usually as many such vectors $\vec{v}$ and corresponding numbers λ as the number

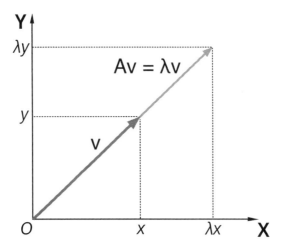

Figure 13.2: Illustration of the geometry of a matrix A multiplying its eigenvector $\vec{v}$, resulting in a vector in the same direction $\lambda\vec{v}$. Figure adapted from Lantonov under CC BY-SA 4.0 via Wikimedia Commons.

of rows or columns of the square matrix A, so a 2-by-2 matrix has two eigenvectors and two eigenvalues, a 5-by-5 matrix has 5 of each, and so forth. One ironclad rule is that there cannot be more distinct eigenvalues than the matrix dimension. Some matrices possess fewer eigenvalues than the matrix dimension; those are said to have a *degenerate* set of eigenvalues, and at least two of the eigenvectors share the same eigenvalue.

The situation with eigenvectors is trickier. For some matrices, any vector is an eigenvector, and others have a limited set of eigenvectors. What is difficult about counting eigenvectors is that an eigenvector is still an eigenvector when multiplied by a constant. You can show that for any matrix, multiplication by a constant is commutative: $cA = Ac$, where A is a matrix and c is a constant. This leads us to the important result that if $\vec{v}$ is an eigenvector with eigenvalue λ, then any scalar multiple $c\vec{v}$ is also an eigenvector with the same eigenvalue. The following demonstrates this algebraically:

$$A \times (c\vec{v}) = cA \times \vec{v} = c\lambda\vec{v} = \lambda(c\vec{v})$$

13.2. EIGENVALUES OF MARKOV MODELS

This shows that when the vector $c\vec{v}$ is multiplied by the matrix A, it results in its being multiplied by the same number λ, so by definition $c\vec{v}$ is an eigenvector. Therefore, an eigenvector $\vec{v}$ is not unique, as any constant multiple $c\vec{v}$ is also an eigenvector. It is more useful to think not of a single eigenvector $\vec{v}$, but of a collection of vectors that can be interconverted by scalar multiplication that are all essentially the same eigenvector. Another way to represent this, if the eigenvector is real, is that an eigenvector is a direction that remains unchanged by multiplication by the matrix, such as direction of the vector $\vec{v}$ in Figure 13.2. As mentioned above, this is true only for real eigenvalues and eigenvectors, since complex eigenvectors cannot be used to define a direction in a real space.

To summarize, the eigenvalues and eigenvectors of a matrix are a set of numbers and a set of vectors (up to scalar multiple), respectively, that describe the action of the matrix as a multiplicative operator on vectors. "Well-behaved" square n-by-n matrices have n distinct eigenvalues and n eigenvectors pointing in distinct directions. In a deep sense, the collection of eigenvectors and eigenvalues defines a matrix A, which is why an older name for them is "characteristic vectors and values."

13.2.2 calculation of eigenvalues on paper

Finding the eigenvalues and eigenvectors analytically (i.e., on paper) is quite laborious even for 3-by-3 or 4-by-4 matrices, and for larger ones there is no analytical solution. In practice, the task is outsourced to a computer, and we will do this using R in section 13.3. Nevertheless, it is useful to go through the process in 2 dimensions to gain an understanding of what is involved. From definition 13.2 of eigenvalues and eigenvectors, the condition can be written in terms of the four elements of a 2-by-2 matrix:

$$\begin{pmatrix} a & b \\ c & d \end{pmatrix} \begin{pmatrix} v_1 \\ v_2 \end{pmatrix} = \begin{pmatrix} av_1 + bv_2 \\ cv_1 + dv_2 \end{pmatrix} = \lambda \begin{pmatrix} v_1 \\ v_2 \end{pmatrix}$$

This is now a system of two linear algebraic equations, which we can solve by substitution. First, let us solve for v_1 in the first row to get

$$v_1 = \frac{-bv_2}{a - \lambda}$$

Then we substitute this into the second equation:

$$\frac{-bcv_2}{a - \lambda} + (d - \lambda)v_2 = 0$$

Since v_2 multiplies both terms, and is not necessarily zero, we require that its multiplicative factor be zero. Doing a little algebra, we obtain the following, known as the *characteristic equation* of the matrix:

$$-bc + (a - \lambda)(d - \lambda) = \lambda^2 - (a + d)\lambda + ad - bc = 0$$

This equation can be simplified by using two quantities defined at the beginning of the section: the sum of the diagonal elements called the trace, $\tau = a + d$, and the determinant, $\Delta = ad - bc$. The quadratic equation has two solutions, dependent solely on τ and Δ:

$$\lambda = \frac{\tau \pm \sqrt{\tau^2 - 4\Delta}}{2} \qquad (13.2)$$

This is the general expression for a 2-by-2 matrix, showing there are two possible eigenvalues. Note that if $\tau^2 - 4\Delta > 0$, the eigenvalues are real, if $\tau^2 - 4\Delta < 0$, they are complex (have both real and imaginary parts), and if $\tau^2 - 4\Delta = 0$, there is only one eigenvalue. The third case is known as degenerate, because two eigenvectors share the same eigenvalue.

Example. Let us take the same matrix we looked at in the previous subsection:

$$A = \begin{pmatrix} 2 & 1 \\ 2 & 3 \end{pmatrix}$$

The trace of this matrix is $\tau = 2 + 3 = 5$, and the determinant is $\Delta = 6 - 2 = 4$. Then by our formula, the eigenvalues are:

$$\lambda = \frac{5 \pm \sqrt{5^2 - 4 \times 4}}{2} = \frac{5 \pm 3}{2} = 4, 1$$

These are the multiples we found in the example above, as expected.

13.2.3 calculation of eigenvectors on paper

The surprising fact is that, as we saw in the last subsection, the eigenvalues of a matrix can be found without knowing its eigenvectors! However, the converse is not true: to find the eigenvectors, one first needs to know the eigenvalues. Given an eigenvalue λ, let us again write down the defining equation of the eigenvector for a generic 2-by-2 matrix:

$$\begin{pmatrix} a & b \\ c & d \end{pmatrix} \begin{pmatrix} v_1 \\ v_2 \end{pmatrix} = \begin{pmatrix} av_1 + bv_2 \\ cv_1 + dv_2 \end{pmatrix} = \lambda \begin{pmatrix} v_1 \\ v_2 \end{pmatrix}$$

This vector equation is equivalent to two algebraic equations:

$$av_1 + bv_2 = \lambda v_1$$
$$cv_1 + dv_2 = \lambda v_2$$

Since we've already found λ by solving the characteristic equation, this is a system of two linear equations with two unknowns (v_1 and v_2). You may remember from advanced algebra that such equations may either have a single solution for each unknown, but sometimes they may have none, or infinitely many solutions. Since there are unknowns on both sides of the equation, we can make both equations be equal to zero:

$$(a - \lambda)v_1 + bv_2 = 0$$
$$cv_1 + (d - \lambda)v_2 = 0$$

So the first equation yields the relationship $v_1 = -v_2 b/(a - \lambda)$, and the second equation is $v_1 = -v_2(d - \lambda)/c$, which we already obtained in the last subsection. We know that these two equations must be the same, since the ratio of v_1 and v_2 is what defines the eigenvector. So we can use either expression to find the eigenvector.

Example. Let us return to the same matrix we looked at in the previous subsection:

$$A = \begin{pmatrix} 2 & 1 \\ 2 & 3 \end{pmatrix}$$

The eigenvalues of the matrix are 1 and 4. Using the expression above, where the element $a = 2$ and $b = 1$, let us find the eigenvector corresponding to the eigenvalue 1:

$$v_1 = -v_2 \times 1/(2-1) = -v_2$$

Therefore the eigenvector is characterized by the first and second elements being negatives of each other. We already saw in the example in subsection 13.2.1 that the vector $(1, -1)$ is such as eigenvector, but it is also true of the vectors $(-1, 1)$, $(-\pi, \pi)$ and $(10^6, -10^6)$. This infinite collection of vectors, all along the same direction, can be described as the eigenvector (or eigendirection) corresponding to the eigenvalue 1.

Repeating this procedure for $\lambda = 4$, we obtain the linear relationship

$$v_1 = -v_2 \times 1/(2-4) = 0.5 v_2$$

Once again, the vector $(2, 1)$ we saw in subsection 13.2.1 is in agreement with our calculation. Other vectors that satisfy this relationship include $(10, 5)$, $(-20, -10)$, and $(-0.4, -0.2)$. This is again a collection of vectors that are all considered the same eigenvector with eigenvalue 4 which are all pointing in the same direction and only differ in their lengths.

Math exercises:

For the following two-state Markov models (a) calculate the eigenvalues of the transition matrix; (b) calculate the corresponding eigenvectors and explain which one corresponds to the stationary distribution; (c) use R to check that each of the eigenvectors obeys definition 13.2 with its corresponding eigenvalue.

Exercise 13.2.1. Use the model in the transition diagram in Figure 10.3a.

Exercise 13.2.2. Use the model in the transition diagram in Figure 10.3b.

Exercise 13.2.3. Use the model in the transition diagram in Figure 10.3c.

13.2. EIGENVALUES OF MARKOV MODELS

Exercise 13.2.4. Use the model in the transition diagram in Figure 10.3d.

Exercise 13.2.5. An ion channel can be in either an open or closed state. If it is open, then it has probability 0.1 of closing in 1 microsecond; if closed, it has probability 0.3 of opening in 1 microsecond.

Exercise 13.2.6. An individual can be either susceptible or infected. The probability of infection for a susceptible person is 0.05 per day, and the probability an infected person becoming susceptible is 0.12 per day.

Exercise 13.2.7. The genotype of an organism can be either normal (wild type) or mutant. Each generation, a wild type individual has probability 0.03 of having a mutant offspring, and a mutant has probability 0.005 of having a wild type offspring.

13.2.4 rate of convergence

Consider a two-state Markov model with the transition matrix M. As we know, the probability distribution vector at time $t + 1$ is the matrix M multiplied by the probability distribution vector at time t:

$$P(t+1) = M \times P(t)$$

Using eigenvectors and eigenvalues, the matrix multiplication (which is difficult) can be turned into multiplication by scalar numbers (which is much simpler). Suppose that the initial probability vector $P(0)$ can be written as a weighted sum (linear combination) of the two eigenvectors of the matrix M, $\vec{v}_1$ and $\vec{v}_2$: $P(0) = c_1\vec{v}_1 + c_2\vec{v}_2$. I will explain exactly how to find the constants c_1 and c_2 a few paragraphs later, but for now, let's go with this. Multiplying the matrix M and this weighted sum (matrix multiplication can be distributed), we get:

$$P(1) = M \times P(0) = M \times (c_1\vec{v}_1 + c_2\vec{v}_2) = c_1 M \times \vec{v}_1 + c_2 M \times \vec{v}_2$$
$$= c_1\lambda_1\vec{v}_1 + c_2\lambda_2\vec{v}_2$$

The last step is due to definition 13.2 of eigenvectors and eigenvalues, which transformed matrix multiplication into multiplication by the corresponding eigenvalues. To see how useful this is, let us propagate the probability vector one more step:

$$P(2) = M \times P(1) = M \times (c_1 \lambda_1 \vec{v}_1 + c_2 \lambda_2 \vec{v}_2) = c_1 \lambda_1^2 \vec{v}_1 + c_2 \lambda_2^2 \vec{v}_2$$

It should be clear that each matrix multiplication results in one additional multiplication of each eigenvector by its eigenvalue, so this allows us to write the general expression for the probability vector any number of time steps t in the future, given the weights c_1 and c_2 of the initial probability vector:

$$P(t) = c_1 \lambda_1^t \vec{v}_1 + c_2 \lambda_2^t \vec{v}_2$$

The constants c_1, c_2 are determined by the initial conditions, while the constants λ_1, λ_2 are the eigenvalues, and the vectors $\vec{v}_1, \vec{v}_2$ are the eigenvectors of the matrix M. This expression is also true for Markov models of any number of states, except that they have as many eigenvalues and eigenvectors as the dimensionality of the transition matrix. This is a hugely important development, because it allows us to predict how quickly the probability vectors converge to the stationary distribution. First, we need to use the following theorem.

Theorem 13.1. (Frobenius) A Markov transition matrix M, characterized by having all nonnegative elements between 0 and 1, and whose columns all sum up to 1, has eigenvalues that are no greater than 1 in absolute value, including at least one eigenvalue equal to 1.

This theorem has an immediate important consequence for the dynamics of the probability vector. According to our formula describing the time evolution of the probability vector, the eigenvalues are raised to the power t, which is the number of time steps. Therefore, for any eigenvalue less than 1, the number λ^t grows smaller and approaches 0 as time goes on. Since Frobenius says that the eigenvalues cannot be greater than 1, the terms in the expression for the probability vector decay, except for the ones equal to 1 or to -1.

13.2. EIGENVALUES OF MARKOV MODELS

The eigenvectors with eigenvalue 1 correspond to the stationary distribution introduced in Chapter 12, and true to their name, they remain unchanged by time, since $1^t = 1$ for all time. The ones with eigenvalue -1 are a strange case, because they oscillate between positive and negative values, without decaying in absolute value.

We now have the skills to answer the following question of practical importance: How quickly does the probability vector approach the stationary distribution? (There may be more than one stationary distribution vector, but that doesn't change the analysis.) The answer depends on how fast the contributions of other, nonstationary eigenvectors decay. If one of them has an eigenvalue of -1, then its contribution never decays—we saw an example of that in the cyclic two-state matrix in section 12.2. If all of the eigenvalues other than the stationary one are less than 1 in absolute value, then all of them decay to zero, but at different rates. The one that decays slowest is the largest of the eigenvalues that are less than one—that is, the second-largest, sometimes called the subdominant eigenvalue. This is the eigenvalue that determines the rate of convergence to the stationary distribution, because it is the "last person standing" of the nonstationary eigenvalues, after the others have all vanished into insignificance.

Example. Consider a Markov model with the following transition matrix M, with eigenvectors and eigenvalues already solved by a computational assistant:

$$M = \begin{pmatrix} 0.8 & 0.1 & 0.1 \\ 0.1 & 0.8 & 0.2 \\ 0.1 & 0.1 & 0.7 \end{pmatrix}$$

$$\lambda_1 = 1 \; \vec{v}_1 = \begin{pmatrix} 1/3 \\ 5/12 \\ 1/4 \end{pmatrix} \quad \lambda_2 = 0.7 \; \vec{v}_2 = \begin{pmatrix} -1 \\ 1 \\ 0 \end{pmatrix} \quad \lambda_3 = 0.6 \; \vec{v}_3 = \begin{pmatrix} 0 \\ -1 \\ 1 \end{pmatrix}$$

Let us compute, using the tools of this section, how the probability distribution vector evolves starting with $P(0) = (7/12, 5/12, 0)$ (I chose this particular initial probability distribution, because it makes the algebra simple, but you can start with any initial vector you want.) The first step is to find what is called the decomposition of the initial probability vector into its three eigenvectors:

$$P(0) = \begin{pmatrix} 7/12 \\ 5/12 \\ 0 \end{pmatrix} = c_1 \begin{pmatrix} 1/3 \\ 5/12 \\ 1/4 \end{pmatrix} + c_2 \begin{pmatrix} -1 \\ 1 \\ 0 \end{pmatrix} + c_3 \begin{pmatrix} 0 \\ -1 \\ 1 \end{pmatrix}$$

It turns out that the desired coefficients are $c_1 = 1$, $c_2 = -1/4$, and $c_3 = -1/4$; you can check yourself that they add up to the initial vector we want. Therefore, after some number of time steps t, the probability distribution vector can be expressed as

$$P(t) = \begin{pmatrix} 1/3 \\ 5/12 \\ 1/4 \end{pmatrix} - \frac{1}{4}0.7^t \begin{pmatrix} -1 \\ 1 \\ 0 \end{pmatrix} - \frac{1}{4}0.6^t \begin{pmatrix} 0 \\ -1 \\ 1 \end{pmatrix}$$

All three eigenvectors, multiplied by their eigenvalues to the power t, are present in the expression, but the third eigenvalue (0.6) decays much faster than the second one (0.7). After 5 time steps, $0.7^5 = 0.168$, while $0.6^5 = 0.078$, so the contribution of the third eigenvector is about half that of the second; after 10 time steps, $0.7^{10} = 0.028$, while $0.6^{10} = 0.006$, so the contribution of the third eigenvector is about one-fifth that of the second; after 20 time steps, $0.7^{20} \approx 8 \times 10^{-4}$, while $0.6^{20} \approx 4 \times 10^{-5}$, so the contribution of the third eigenvector is about one-twentieth that of the second. The trend is clear: although 0.6 and 0.7 are not that different, raising them to higher powers makes the ratio between them get smaller, until the contribution of the smaller of the two eigenvalues is negligible. However you define that term—less than 1%? less than 0.01%?—eventually the smaller eigenvalue will reach that level of insignificance. This illustrates why the rate of convergence to the stationary distribution $(1/3, 5/12, 1/4)$ is determined by the eigenvalue 0.7, and the smaller eigenvalue can be neglected.

13.3 Matrix diagonalization in R

R has a standard set of functions to handle linear algebra computations. One of the most important of those functions is the calculation of eigenvectors and eigenvalues, also known as *diagonalization*

13.3. MATRIX DIAGONALIZATION IN R

of a matrix (for reasons you'll understand if you take a proper linear algebra course in the future, as I strongly recommend). The function to calculate the special numbers and vectors is `eigen()`, and the name of the matrix goes between the parentheses. The function returns a data frame as its output, with $values and $vectors storing the eigenvalues and the eigenvectors, respectively. Here is a script to define the matrix analyzed in the examples in section 13.2 and then calculate and print out its eigenvalues and eigenvectors:

```
test.matrix <- matrix(c(2, 2, 1, 3), nrow = 2)
print(test.matrix)
```

```
##      [,1] [,2]
## [1,]    2    1
## [2,]    2    3
```

```
result <- eigen(test.matrix)
print(result$values)
```

```
## [1] 4 1
```

```
print(result$vectors)
```

```
##            [,1]       [,2]
## [1,] -0.4472136 -0.7071068
## [2,] -0.8944272  0.7071068
```

The resulting eigenvalues are 4 and 1, just as we computed above. However, the situation with eigenvectors is trickier. In the R output they are presented as column vectors, and you may notice that they have the same ratio of elements as the calculated eigenvectors, but the numbers may appear strange. As we discussed, the eigenvector for a particular eigenvalue can be multiplied by any constant and still be a valid eigenvector. R (and other computational tools) thus has a choice about which eigenvector to output, and it typically prefers to *normalize* its eigenvectors by making sure their length (Euclidean norm) is equal to 1. You may observe that if you square the two elements of one of the eigenvectors in the output and add them up, the result will be 1. This is convenient for some purposes, but doesn't make for clean looking vectors in terms of the elements. But since we can multiply (or divide) the eigenvector by any constant, we can choose to make it look cleaner, for instance, by making sure one of its elements is equal to 1. The next script illustrates how to make that happen by dividing each eigenvector by the value of its first element, which results in the same form of eigenvectors that we wrote down in the analytical solution:

```
eigen1 <- result$vectors[, 1]/result$vectors[1, 1]
print(eigen1)
```

```
## [1] 1 2
```

```
eigen2 <- result$vectors[, 2]/result$vectors[1, 2]
print(eigen2)
```

```
## [1]  1 -1
```

Let us now analyze a transition matrix, for example, one that we investigated in subsection 13.2.4. This is how I obtained the eigenvalues and eigenvectors that were used to make predictions about the evolution of the probability vector. One important issue here is how best to normalize the eigenvectors. The eigenvector corresponding the eigenvalue 1 is (a multiple of) the stationary distribution vector, but to make it a probability distribution vector, its elements must add up to 1. The way to ensure this property is to divide the eigenvector by the sum of its elements. The two other eigenvectors are not probability vectors (in fact, they both contain negative elements), so it doesn't make sense to normalize them the same way. Any normalization of these vectors is arbitrary, so I choose to divide each by the value of one of its elements. Notice one more strange thing: the two nonstationary eigenvectors each contain a very small number on the order of 10^{-16}. You may remember from section 1.3 that this is the limit of precision for storing numbers in R, so these values are not real, they are artifacts of rounding errors, and instead should be replaced with zeros.

13.3. MATRIX DIAGONALIZATION IN R

```
trans.matrix <- matrix(c(0.8, 0.1, 0.1, 0.1, 0.8, 0.1,
    0.1, 0.2, 0.7), nrow = 3)
print(trans.matrix)
```

```
##      [,1] [,2] [,3]
## [1,]  0.8  0.1  0.1
## [2,]  0.1  0.8  0.2
## [3,]  0.1  0.1  0.7
```

```
result <- eigen(trans.matrix)
print(result$values)
```

```
## [1] 1.0 0.7 0.6
```

```
print(result$vectors)
```

```
##           [,1]           [,2]           [,3]
## [1,] 0.5656854 -7.071068e-01 -6.864310e-16
## [2,] 0.7071068  7.071068e-01 -7.071068e-01
## [3,] 0.4242641  3.330669e-16  7.071068e-01
```

```
eigen1 <- result$vectors[, 1]/sum(result$vectors[,
    1])
print(eigen1)
```

```
## [1] 0.3333333 0.4166667 0.2500000
```

```
eigen2 <- result$vectors[, 2]/result$vectors[1, 2]
print(eigen2)
```

```
## [1]  1.000000e+00 -1.000000e+00 -4.710277e-16
```

```
eigen3 <- result$vectors[, 3]/result$vectors[2, 3]
print(eigen3)
```

```
## [1]  9.7076e-16  1.0000e+00 -1.0000e+00
```

We now have the tools to calculate the eigenvalues and eigenvectors of transition matrices, and we can use them to predict two things: (1) the stationary distribution, which is the properly normalized eigenvector with eigenvalue 1, and (2) how quickly probability vectors converge to the stationary distribution, which depends on the second-largest eigenvalue. The first task was already accomplished in the script above.

The second question needs clarification: what does it mean to "converge," that is, how close does the probability distribution need to be to the stationary one, before we can say it has converged? The answer is necessarily arbitrary, because a probability vector never actually reaches the stationary distribution unless the initial vector is stationary. Distances between vectors are typically measured using the standard Euclidean definition of distance (the square root of the sum of the squares of differences of the elements). The following R script propagates an initial probability vector for 30 time steps, and at each step calculates the distance between the probability vector and the stationary vector (which we calculated above).

```
nstep <- 30   # set number of time steps
stat.vec <- eigen1   # set the stationary distribution
prob.vec <- c(1, 0, 0)
# set the initial probability vector
dist.vec <- rep(0, nstep)   # initialize distance vector
dist.vec[1] <- sqrt(sum((prob.vec - stat.vec)^2))
for (i in 2:nstep) {
    # propagate the probability vector
    prob.vec <- trans.matrix %*% prob.vec
    dist.vec[i] <- sqrt(sum((prob.vec - stat.vec)^2))
}
```

The distance to the stationary vector as a function of time is plotted in Figure 13.3 along with the exponential decay of the second-largest eigenvalue 0.7^t over the same number of time steps. While initially the two are not identical, over time the contribution of the smaller eigenvalue becomes insignificant, and the second-largest eigenvalue describes the approach of the probability vector to the stationary distribution. The effect of choosing a different

13.4. MOLECULAR EVOLUTION

initial probability vector is not very large, with the exception of the case when the initial vector is exactly the stationary distribution (in which case the distance is zero initially and remains zero) or exactly the eigenvector with the smaller eigenvalue, in which case the smallest eigenvalue governs the convergence.

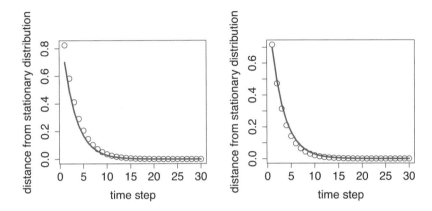

Figure 13.3: Decay of the distance from the stationary distribution (circles) and the exponential decay of the second-largest eigenvalue (solid curves), plotted for two different initial probability distributions.

13.4 Molecular evolution

13.4.1 Jukes-Cantor model

Substitution mutations in DNA sequences can be modeled as a Markov process, where each base in the sequence mutates independently of others with a transition matrix M. Let the bases A, G, C, T correspond to states 1 through 4, respectively. A classic model for base substitution from one generation to the next is based on the assumptions that all substitution mutations are equally likely and that the fraction a of the sequence will be substituted each generation. This was proposed by Jukes and Cantor (1969) to calculate the rate of divergence of DNA (or protein) sequences from their common ancestor. The model is illustrated as a transition diagram in Figure 13.4, with the four letters representing the states of a particular site in a DNA sequence, and all transition probabilities equal to a.

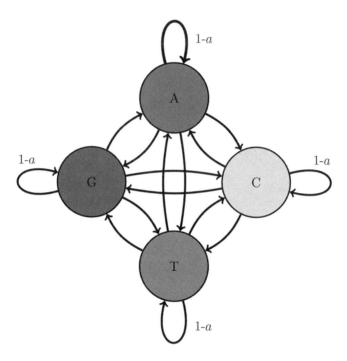

Figure 13.4: Transition diagram for a four-state molecular evolution model for one letter in a DNA sequence. The mutation rate a is the probability that the letter is replaced by a different one over one generation. The transition probabilities between individual letters are $a/3$ (not labeled).

Then the probability of any particular transition, say from T to C, is $a/3$, while the probability of not having a substitution is equal to $1 - a$. This is known as the Jukes-Cantor model, and it predicts that the fraction of letters in a sequence at generation $t+1$ depends on the distribution in generation t as follows:

$$\begin{pmatrix} P_A(t+1) \\ P_G(t+1) \\ P_C(t+1) \\ P_T(t+1) \end{pmatrix} = \begin{pmatrix} 1-a & a/3 & a/3 & a/3 \\ a/3 & 1-a & a/3 & a/3 \\ a/3 & a/3 & 1-a & a/3 \\ a/3 & a/3 & a/3 & 1-a \end{pmatrix} \begin{pmatrix} P_A(t) \\ P_G(t) \\ P_C(t) \\ P_T(t) \end{pmatrix}$$

13.4. MOLECULAR EVOLUTION

This model is very simple. It only considers substitutions, although other mutations are possible (e.g., insertions and deletions) although they are typically more disruptive and thus more rare, and it treats all substitutions as equally likely, which is not empirically true. The benefit is that the number a is the only parameter in this model, which represents the mutation rate at each site per generation. This makes is easy to compute the eigenvectors and eigenvalues of the model in general. It turns out that the four eigenvectors do not depend on the parameter a, only the eigenvalues do:

$$\begin{pmatrix} 1/4 \\ 1/4 \\ 1/4 \\ 1/4 \end{pmatrix} \lambda = 1; \quad \begin{pmatrix} 1/4 \\ -1/4 \\ 1/4 \\ -1/4 \end{pmatrix} \lambda = 1 - \frac{4}{3}a$$

$$\begin{pmatrix} 1/4 \\ -1/4 \\ -1/4 \\ 1/4 \end{pmatrix} \lambda = 1 - \frac{4}{3}a; \quad \begin{pmatrix} 1/4 \\ -1/4 \\ 1/4 \\ -1/4 \end{pmatrix} \lambda = 1 - \frac{4}{3}a$$

Notice two things: first, the first eigenvector is the equilibrium distribution and has the same frequencies for all four bases. Second, the three eigenvectors with eigenvalues smaller than 1 have negative entries, so they cannot be probability distributions themselves (although as linear combinations with the first one, they may be, depending on the coefficients).

Computationally, this model allows us to predict the time evolution of the distribution of bases in a DNA sequence for any given initial distribution, by using repeated matrix multiplication as above. Figure 13.5 shows the probability distribution of nucleotides based on the Jukes-Cantor model starting with the nucleotide A for two different mutation rates, propagated for different lengths of time. It is evident that for a faster substitution rate the approach to the equilibrium distribution is faster. This demonstrates how the second-largest eigenvalue of the transition matrix determines the speed of convergence to the equilibrium distribution, as we postulated in section 13.2.

Thus, the Jukes-Cantor model provides a prediction of the time-dependent evolution of the probability distribution of each letter,

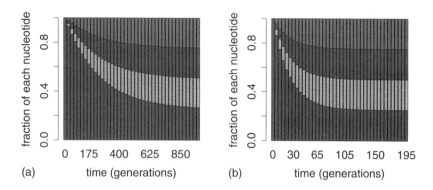

Figure 13.5: Evolution of probability vectors in the Jukes-Cantor model with bar graphs showing proportion of letters A, G, T, C respectively: (a) 1000 generations with mutation rate $a = 0.001$; (b) 200 generations with substitution rate $a = 0.01$.

starting with an initial distribution. In reality, we would like to answer the following question: Given two DNA sequences in the present (e.g., from different species), what is the length of time they spent evolving from a common ancestor?

13.4.2 time since divergence

To do this, we need to some preliminary work. The first step is to compute the probability that a letter at a particular site remains unchanged after t generations. Because all the nucleotides are equivalent in the Jukes-Cantor model, this is the same as finding the probability that a nucleotide in state A remains in A after t generations. As we saw in section 13.2, we can calculate the frequency distribution after t time steps by using the decomposition of the initial probability vector $\vec{P}(0)$ into a weighted sum of the eigenvectors. We take $\vec{P}_0 = (1, 0, 0, 0)$ (i.e., the initial state is A), and this can be written as a sum of the four eigenvectors of the matrix M:

13.4. MOLECULAR EVOLUTION

$$\begin{pmatrix} 1 \\ 0 \\ 0 \\ 0 \end{pmatrix} = \begin{pmatrix} 1/4 \\ 1/4 \\ 1/4 \\ 1/4 \end{pmatrix} + \begin{pmatrix} 1/4 \\ -1/4 \\ 1/4 \\ -1/4 \end{pmatrix} + \begin{pmatrix} 1/4 \\ -1/4 \\ -1/4 \\ 1/4 \end{pmatrix} + \begin{pmatrix} 1/4 \\ -1/4 \\ 1/4 \\ -1/4 \end{pmatrix}$$

Therefore, the transition matrix M can be applied to each eigenvector separately, and each matrix multiplication is a multiplication by the appropriate eigenvalue. Thus,

$$P(t) = M^t P(0) =$$

$$= 1^t \begin{pmatrix} 1/4 \\ 1/4 \\ 1/4 \\ 1/4 \end{pmatrix} + (1-\tfrac{4}{3}a)^t \left(\begin{pmatrix} 1/4 \\ -1/4 \\ 1/4 \\ -1/4 \end{pmatrix} + \begin{pmatrix} 1/4 \\ -1/4 \\ -1/4 \\ 1/4 \end{pmatrix} + \begin{pmatrix} 1/4 \\ -1/4 \\ 1/4 \\ -1/4 \end{pmatrix} \right)$$

The first element of $P(t)$ is the probability of a nucleotide remaining A after t generations, and it is $P_A(t) = 1/4 + 3/4(1-\tfrac{4}{3}a)^t$. For $t = 0$, the probability is 1, as it should be, and as $t \to \infty$, $P_A(t) \to 1/4$, since this is the equilibrium probability distribution. Note that the expression is the same for all the other letters, so we have found the expression for any nucleotide remaining the same after t generations.

Now let us get to the question of calculating the time elapsed since two sequences have evolved from a common ancestor. Denote by m the fraction of sites in two aligned sequences with different letters, and q is the probability of a nucleotide remaining the same, which is the given by the expression for $P_A(t)$. Thus $m = 1 - q = 3/4 - 3/4(1-\tfrac{4}{3}a)^t$. This can be solved for t:

$$t = \frac{\log(1-\tfrac{4}{3}m)}{\log(1-\tfrac{4}{3}a)}$$

13.4.3 calculation of phylogenetic distance

However, if we do not know the mutation rate a—and until recently, it was rarely known with any precision—this formula is of limited practical use. Jukes and Cantor neatly finessed the problem by calculating the *phylogenetic distance* between the two sequences, which is defined as $d = at$, or the mean number of substitutions that occurred per nucleotide during t generations, with mutation rate a (substitutions per nucleotide per generation). Note that this distance is not directly measurable from the fraction of different nucleotides in the two sequences, because it counts all substitutions, including those reversing an earlier mutation and so causing the sequence to revert to its initial letter.

Now, let us assume a is small, as is usually the case; as you recall from section 3.1, for humans, the rate of substitutions per generation per nucleotide is about 10^{-8}. By Taylor expansion of the logarithm around 1, $\log(1 - \frac{4}{3}a) \approx -\frac{4}{3}a$. Using the formula for t from above with this approximation, we find the Jukes-Cantor phylogenetic distance to be

$$d_{JC} = \frac{\log(1 - \frac{4}{3}m)}{-\frac{4}{3}a} a = -\frac{3}{4} \log(1 - \frac{4}{3}m)$$

This formula has the correct behavior in the two limits: when $m = 0$, $d_{JC} = 0$ (identical sequences have zero distance), and when $m \to 3/4$, $d_{JC} \to \infty$, since $3/4$ is the maximum possible fraction of differences under the Jukes-Cantor model. Thus, we have obtained an analytic formula for the phylogenetic distance based on the fraction of differences between two homologous sequences.

A useful discussion of the details of connecting the Jukes-Cantor distance to phylogenetic distances can be found on the treethinkers blog (Thomson 2013), describing several possible adjustments that can be made to the formula. It can then be used to calculate a *distance matrix* that contains distances between all pairs from a collection of sequences. There are algorithms that can generate phylogenetic trees based on a distance matrix, which minimize the total phylogenetic distance of all the branches. This is one approach that biologists use to infer evolutionary relationships from molecular sequence data, but there are many others, including maximum-likelihood and parsimony models.

13.4.4 divergence of human and chimp genomes

The human genome sequence was reported in 2001, and the genome of our closest living interspecies relative, the chimpanzee, was published in 2005 (Chimpanzee Sequencing and Analysis Consortium

13.4. MOLECULAR EVOLUTION

2005). Comparison between the two genomes has allowed us to peer into the evolutionary history of the two lineages. Although our genomes are quite similar, several large genome changes separate us and chimps. One particularly big difference is the number of chromosomes: the human genome is organized into 23 chromosomes, while chimps have 24. This suggests two options: either the common ancestor had 23 chromosomes and one of them split in the chimpanzee lineage, or the common ancestor has 24 and and two of them merged in the human lineage. Determining the complete sequences has answered that questions decisively: the human chromosome 2 is a result of a fusion of two ancestral chromosomes. This is supported by multiple observations, in particular, the two halves of human chromosome 2 can be matched to contiguous sequences in two separate chimp chromosomes, and the presence of telomere sequences (which form caps on the ends of chromosomes) in the middle of human chromosome 2. The entire story of great ape chromosomal evolution is still in the process of being investigated. One recent study that examines the history of gorilla and orangutan chromosomal modification as well is Ventura et al. (2012).

Besides the large-scale changes, many point mutations have accumulated since the split of the two lineages. Large portions of the two genomes can be directly aligned, and the differences can be measured as the fractions of letters that do not match in a particular region. In the original report on the chimp genome (Chimpanzee Sequencing and Analysis Consortium 2005), the authors calculated the divergence (mean fraction of different letters in an aligned sequence) to be 1.23%. The distribution of these divergences is shown in Figure 13.6. This information allows us to calculate the likely time since the split of the two lineages to be about 6–7 million years. The model used to make that calculation is more sophisticated than the simple Jukes-Cantor substitution model and uses a maximum-likelihood approach.

Discussion questions:

The following questions refer to the paper "Initial Sequence of the Chimpanzee Genome and Comparison with the Human Genome" (Chimpanzee Sequencing and Analysis Consortium 2005).

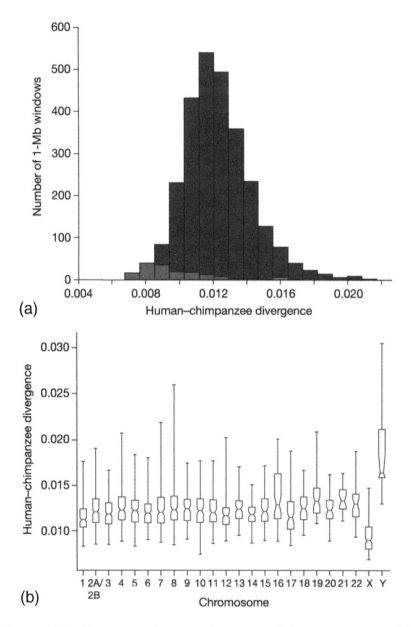

Figure 13.6: Divergence between human and chimp genomes: (a) histogram of divergence values for 1 Mb(million bases) long segments of the genome (■ indicates segments from autosomes, ▩ from the X chromosome; and □ from the Y chromosome); (b) divergence distribution for different chromosomes; figure from Chimpanzee Sequencing and Analysis Consortium (2005), used by permission.

13.5. COMPUTATIONAL PROJECTS

Discussion 13.4.1. Describe some challenges in comparing genomes between two different species.

Discussion 13.4.2. Speculate on the biological reasons for the disparities in the substitution rates in different chromosomes and in the distal parts (those closer to the ends of the chromosome) compared to the proximal parts (closer to the center).

Discussion 13.4.3. What are the differences in the transposable elements (SINES and LINES) between the two genomes? Are there any explanations offered for this observation?

Discussion 13.4.4. Comment on the meaning of the K_A/K_S ratio for studying the effect of natural selection.

13.5 Computational projects

13.5.1 eigenvalues of a two-state model

In this section we return to the two-state SI model (see section 10.5) with the same transition probabilities (per day): 0.04 (from S to I) and 0.12 (from I to S).

Tasks

1. Write down the transition matrix for the model, and assign it to a matrix in R. Find the eigenvalues and eigenvectors of the matrix, and predict the stationary distribution vector.

2. Calculate the probability distribution vectors for 100 days using several initial probability vectors of your choice, and plot the probability vectors over time. How different are the probability distributions after 100 days? Explain how the plot of probability distribution over time is connected to the eigenvalues and eigenvectors you found in task 1.

3. Use the script in section 13.3 to compute the distance from the probability vector to the stationary distribution, and plot it over time. Describe its functional form and its relationship to the eigenvalues and eigenvectors you found in task 1.

4. Change the infection rate (transition from S to I) to 0.2, and assign the new values to the transition matrix, Find the eigenvalues and eigenvectors of the matrix, and predict the new stationary distribution vector.

5. Repeat the computations in task 2 with the new transition matrix and several initial probability vectors, and plot the probability distributions over time. Are the probability distributions based on the two initial vectors different from each other? How is their time evolution different from the distributions in task 4? Explain how the plot of probability distribution over time is connected to the eigenvalues and eigenvectors you found in task 4.

6. Compute the distance from the probability vector to the stationary distribution, and plot it over time. Describe its functional form and its relationship to the eigenvalues and eigenvectors you found in task 4.

13.5.2 eigenvalues of a three-state model

In this section we return to the three-state model of nicotinic acetylcholine receptor (nAChR) discussed in Chapters 10–12. As before, set the transition probabilities to the following: 0.04 (from R to C), 0.07 (from C to R), 0.12 (from C to O), and 0.02 (from O to C); the other transition probabilities are 0.

Tasks

1. Write down the transition matrix for the model, and assign it to a matrix in R. Find the eigenvalues and eigenvectors of the matrix, and predict the stationary distribution vector.

2. Suppose that the ion channel starts out in the state R. Write a script to propagate the probability vector for any given length of time, and plot the probability vector over 1000 microseconds. Does the distribution converge to an equilibrium and if so, over how many time steps? Now start with an initial probability vector of your choice, and plot the probability vector

13.5. COMPUTATIONAL PROJECTS

over 1000 microseconds. Explain how these plots of probability distribution over time are connected to the eigenvalues and eigenvectors you found in task 1.

3. Compute the distance from the probability vector to the stationary distribution, and plot it over time. Describe its functional form and its relationship to the eigenvalues and eigenvectors you found in task 1.

4. Change the opening rate (transition from C to O) to 0.02, and assign the new values to the transition matrix. Find the eigenvalues and eigenvectors of the matrix, and predict the new stationary distribution vector.

5. Repeat the computations in task 2 with the new transition matrix from task 4, and plot the probability vector over 1000 microseconds. Does the distribution converge to an equilibrium and if so, over how many time steps? Now start with an initial probability vector of your choice, and plot the probability vector over 1000 microseconds. Explain how these plots of probability distribution over time are connected to the eigenvalues and eigenvectors you found in task 4.

6. Compute the distance from the probability vector to the stationary distribution, and plot it over time. Describe its functional form and its relationship to the eigenvalues and eigenvectors you found in task 4.

13.5.3 analysis of the Jukes-Cantor model

This project is challenging, because it requires you to implement a simulation with minimal guidance and run numerical experiments.

1. Write a function to generate a mutant sequence of nucleotides (letters A, T, G, C) based on the Jukes-Cantor model. It should take an initial (ancestral) sequence, the mutation rate, and the number of generations, and return the "evolved" sequence.

2. Use your function to generate two sequences starting with the same ancestral sequence; that is, simulate two sequences evolving separately after a split. Calculate the fraction of different letters between the two sequences, and compare them to the predicted number of generations based on the fraction of difference from the Jukes-Cantor model.

3. Repeat the experiment many times, and report the statistics of the fraction of differences for a given mutation rate for a certain number of generations. Vary the number of generations, and compare the mean value of the expected number of generations to the time in the Jukes-Cantor model.

Part IV
Variables that change with time

Chapter 14

Linear difference equations

> *We're captive on the carousel of time*
> *We can't return we can only look behind*
> *From where we came*
> *And go round and round and round*
> *In the circle game*
>
> —Joni Mitchell, *The Circle Game*

All living things change over time, and this evolution can be quantitatively measured and analyzed. Mathematics makes use of equations to define models that change with time, known as *dynamical systems*. This part discusses how to construct models that describe the time-dependent behavior of some measurable quantity in life sciences. Numerous fields of biology use such models, and in particular, we will consider changes in population size, the progress of biochemical reactions, the spread of infectious disease, and the spikes of membrane potentials in neurons as some of the main examples of biological dynamical systems.

Many processes in living things happen regularly, repeating with a fairly constant time period. One common example is the reproductive cycle in species that reproduce periodically, whether once a year, or once an hour, like certain bacteria that divide at a relatively constant rate under favorable conditions. Other periodic phenomena include circadian (daily) cycles in physiology, contractions of the heart muscle, and waves of neural activity. For these processes theoretical biologists use models with *discrete time*, in which the time variable is restricted to the integers. For instance, it is natural to count the generations in whole numbers when modeling population growth.

This chapter is devoted to analyzing dynamical systems in which time is measured in discrete steps. In this chapter you will learn to do the following:

1. write down discrete-time (difference) equations based on stated assumptions,

2. find analytic solutions of linear difference equations,

3. define functions in R, and

4. compute numerical solutions of difference equations.

14.1 Discrete-time population models

Let us construct our first models of biological systems. We start by considering a population of some species, with the goal of tracking its growth or decay over time. The variable of interest is the number of individuals in the population, which we call N. This is called the dependent variable, since its value changes depending on time; it would make no sense to say that time changes depending on the population size. Throughout the study of dynamical systems, we will denote the independent variable of time by t. To denote the population size at time t we can write $N(t)$ but sometimes use N_t.

14.1.1 static population

To describe the dynamics, we need to write down a rule for how the population changes. Consider the simplest case, in which the population stays the same for all time. (Maybe it is a pile of rocks?) Then the following equation describes this situation:

$$N(t+1) = N(t)$$

14.1. DISCRETE-TIME POPULATION MODELS

This equation mandates that the population at the next time step be the same as at the present time t. This type of equation is generally called a *difference equation*, because it can be written as a difference between the values at the two different times:

$$N(t+1) - N(t) = 0$$

This version of the model illustrates that a difference equation at its core describes the *increments* of N from one time step to the next. In this case, the increments are always 0, which makes it plain that the population does not change from one time step to the next.

14.1.2 exponential population growth

Let us consider a more interesting situation: a colony of dividing bacteria, such as *Escherichia coli*, shown in Figure 14.1. Assume that each bacterial cell divides and produces two daughter cells at fixed intervals of time, and let us further suppose that bacteria never die. Essentially, we are assuming a population of immortal bacteria with clocks. This means that after each cell division, the population size doubles. As before, we denote the number of cells in each generation by $N(t)$ and obtain the equation describing each successive generation:

$$N(t+1) = 2N(t)$$

It can also be written in the difference form:

$$N(t+1) - N(t) = N(t)$$

The increment in population size is determined by the current population size, so the population in this model is forever growing. This type of behavior is termed *exponential growth*, which we investigate further in section 14.2.

14.1.3 example with birth and death

Suppose that a type of fish lives to reproduce only once after a period of maturation, after which the adults die. In this simple scenario, half the population is female; a female always lays 1000

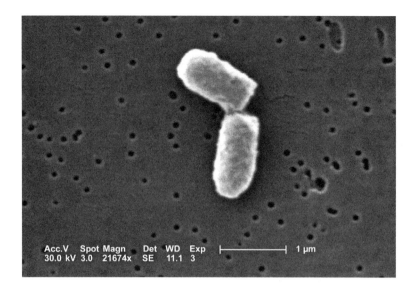

Figure 14.1: Scanning electron micrograph of a dividing *Escherichia coli* bacteria. Image by Evangeline Sowers and Janice Haney Carr (CDC) in public domain via Wikimedia Commons.

eggs; and of those, 1% survive to maturity and reproduce. Let us set up the model for the population growth of this idealized fish population. The general idea, as before, is to relate the population size at the next time step $N(t+1)$ to the population at the present time $N(t)$.

Let us tabulate both the increases and the decreases in the population size. We have $N(t)$ fish at the present time, but we know they all die after reproducing, so there is a decrease of $N(t)$ in the population. Since half of the population is female, the number of new offspring produced by $N(t)$ fish is $500N(t)$. Of those, only 1% survive to maturity (the next time step), and the other 99% ($495N(t)$) die. We can add all the terms together to obtain the following difference equation:

$$N(t+1) = N(t) - N(t) + 500N(t) - 495N(t) = 5N(t)$$

The number 500 in the expression is the *birth rate* of the population per individual, and the negative terms add up to the *death rate* of 496 per individual. We can rewrite the equation in difference form:

$$N(t+1) - N(t) = 4N(t)$$

This expression again generates growth in the population, because the birth rate outweighs the death rate (Allman and Rhodes 2003).

14.1.4 dimensions of birth and death rates

As discussed in section 2.1, the dimensions of quantities in a model have to satisfy the rules of *dimensional analysis*. In the case of population models, the birth and death rates measure the number of individuals that are born (or die) within a reproductive cycle for every individual at the present time. Their dimensions must be such that the terms in the equation all match:

$$[N(t+1) - N(t)] = [\,population\,] = [r][N(t)] = [r] \times [\,population\,]$$

This implies that $[r]$ is algebraically dimensionless. However, the meaning of r is the rate of change of population over one (generation) time step: r is the birth or death rate of the population *per generation*, and therefore, when such rates are measured, they are reported with units of inverse time (e.g., number of offspring per year).

We can now write a general difference equation for any population with constant birth and death rates. This allows us to substitute arbitrary values of the birth and death rates to model different biological situations. Suppose that a population has a birth rate of b per individual, and a death rate d per individual. Then the general model of the population size is:

$$N(t+1) = (1 + b - d)N(t) \tag{14.1}$$

The general equation also allows us to check the dimensions of birth and death rates, especially as written in the incremental form: $N(t+1) - N(t) = (b-d)N(t)$. The change in population rate over one reproductive cycle is given by the current population size multiplied by the difference of birth and death rates, which as we saw are algebraically dimensionless. The right-hand side of the equation has the dimensions of population size, matching the difference on the left-hand side (Edelstein-Keshet 2005).

14.2 Solutions of linear difference models

14.2.1 simple linear models

Having set up the difference equation models, we would naturally like to solve them to find out how the dependent variable, such as population size, varies over time. A solution may be *analytical*, meaning that it can be written as a formula, or *numerical*, in which case it is generated by a computer in the form of a sequence of values of the dependent variable over a period of time. In this section we find some simple analytic solutions and learn to analyze the behavior of difference equations that cannot be solved exactly.

Definition 14.1. A function $N(t)$ is a *solution* (over some time $a < t < b$) of a difference equation $N(t+1) = f(N(t))$ if it satisfies that equation (over $a < t < b$).

For instance, let us take our first model of the static population, $N(t+1) = N(t)$. Any constant function is a solution, for example, $N(t) = 0$, or $N(t) = 10$. There are actually as many solutions as there are numbers, that is, infinitely many! To specify exactly what happens in the model, we need to specify the size of the population at some point, usually, at the "beginning of time," $t = 0$. This is called the *initial condition* for the model, and for a well-behaved difference equation, it is enough to determine a unique solution. For the static model, specifying the initial condition is the same as specifying the population size for all time.

Now let us look at the general model of population growth with constant birth and death rates. Equation 14.1 showed that these can be written in the form $N(t+1) = (1+b-d)N(t)$. To simplify, let us combine the numbers into one growth parameter, $r = 1+b-d$, and write down the general equation for population growth with constant growth rate:

$$N(t+1) = rN(t)$$

To find the solution, consider a specific example, where we start with the initial population size $N_0 = 1$, and the growth rate $r = 2$. The sequence of population sizes is 1, 2, 4, 8, 16, and so forth. This is described by the formula $N(t) = 2^t$.

14.2. SOLUTIONS OF LINEAR DIFFERENCE MODELS

In the general case, for each time step the solution is multiplied by r, so the solution has the same exponential form. The initial condition N_0 is a multiplicative constant in the solution, and one can verify that when $t = 0$, the solution matches the initial value:

$$N(t) = r^t N_0 \qquad (14.2)$$

Pause and consider this remarkable formula. No matter what birth and death parameters are selected, this solution predicts the population size at any point in time t.

To verify that the formula for $N(t)$ is actually a solution in the sense of definition 14.1, we need to check that it actually satisfies the difference equation for all t, not just a few time steps. This can be done algebraically by plugging in $N(t+1)$ into the left-hand side of the dynamic model and $N(t)$ into the right-hand side and checking whether they match. For $N(t)$ given by equation 14.2, $N(t+1) = r^{t+1} N_0$, and thus the dynamic model becomes

$$r^{t+1} N_0 = r \times r^t N_0$$

Since the two sides match, the solution is correct.

14.2.2 models with a constant term

Now let us consider a dynamic model that combines two different rates: a proportional rate (rN) and a constant rate that does not depend on the value of the variable N. We can write such a generic model as

$$N(t+1) = rN(t) + a$$

The right-hand side of this equation is a linear function of N, so this is a linear difference equation with a constant term. What function $N(t)$ satisfies it? One can quickly check that the same solution $N(t) = r^t N_0$ does not work because of the pesky constant term a:

$$r^{t+1} N_0 \neq r \times r^t N_0 + a$$

To solve it, we need to try a different form: specifically, an exponential with an added constant. The exponential can be reasonably surmised to have base r as before, and then leave the two constants as unknown: $N(t) = c_1 r^t + c_2$. To figure out whether this is a solution, plug it into the linear difference equation above, and check whether a choice of constants can make the two sides agree:

$$N(t+1) = c_1 r^{t+1} + c_2 = rN(t) + a = rc_1 r^t + rc_2 + a$$

This equation has the same term $c_1 r^{t+1}$ on both sides, so they can be subtracted out. The remaining equation involves only c_2, and its solution is $c_2 = a/(1-r)$. Therefore, the general solution of this linear difference equation is the following expression, which is determined from the initial value by plugging $t = 0$ and solving for c:

$$N(t) = cr^t + \frac{a}{1-r} \tag{14.3}$$

Example. Take the difference equation $N(t+1) = 0.5N(t) + 40$ with initial value $N(0) = 100$. The solution, according equation 14.3 is $N(t) = c0.5^t + 80$. At $N(0) = 100 = c + 80$, so $c = 20$. Then the complete solution is $N(t) = 20 \times 0.5^t + 80$. To check that this actually works, plug this solution back into the difference equation:

$$N(t+1) = 20 \times 0.5^{t+1} + 80 = 0.5 \times (20 \times 0.5^t + 80) + 40$$
$$= 20 \times 0.5^{t+1} + 80$$

The equation is satisfied, and therefore the solution is correct.

14.3 Population growth and decline

The parameter r can assume different values, depending on the birth and death rates. If the birth rate is greater than the death rate, $r > 1$, and if it is the other way around, $r < 1$. Note that for a realistic biological population, the death rate is limited by the number of individuals present in the population. The maximum number of individuals at any time is $N(t) + bN(t)$, so this means that $d \leq b + 1$. Therefore, for a biological population, $r \geq 0$.

The solutions in equations 14.2 and 14.3 are exponential functions which, as we saw in section 2.2, have a limited menu of behaviors, depending on the value of r. If $r > 1$, multiplication by r increases the size of the population, so the solution $N(t)$ will grow.

14.3. POPULATION GROWTH AND DECLINE

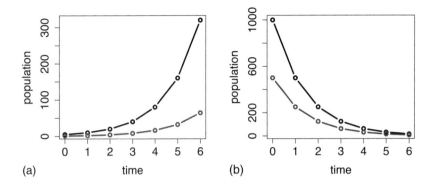

Figure 14.2: Plots of solutions of linear difference equations: (a) $N(t+1) = 2N(t)$ with initial values $N(0) = 5$ and $N(0) = 1$; (b) $N(t+1) = 0.5N(t)$ with initial values $N(0) = 1000$ and $N(0) = 500$.

If $r < 1$, multiplication by r decreases the size of the population, so the solution $N(t)$ will decay (see Figure 14.2). Finally, if $r = 1$, multiplication by r leaves the population size unchanged, as in the pile of rocks model. Here is the complete classification of the behavior of population models with constant birth and death rates (assuming $r > 0$):

- $r > 1$: $N(t)$ grows without bound
- $r < 1$: $N(t)$ decays to the constant $a/(1-r)$
- $r = 1$: the absolute value of $N(t)$ remains constant

Thus there are only two options for the solutions of linear difference equations: ever-faster growth or decay to zero or another constant value. The exponential growth of populations is also known as *Malthusian* after the early population modeler Thomas Malthus. He used a simple population model with constant growth rate to predict demographic disaster due to an exponentially increasing population outstripping the growth in food production. In fact, the human population has not been growing with a constant birth rate, and food production has (so far) kept pace with population size,

316 CHAPTER 14. LINEAR DIFFERENCE EQUATIONS

illustrating yet again that mathematical models are only as good as the assumptions that underlie them.

Math exercises:

For the following scenarios for a population (a) construct a dynamic model by writing down a difference equation both in the updating function form $(N(t+1) = f(N(t)))$ and the increment form $(N(t+1) - N(t) = g(N(t)))$, and specify the time step; (b) find the solution of the linear difference equation with a generic initial value, and check that it satisfies the difference equation by plugging the solution into your equation (in either form); (c) plug in the given initial value, and predict the future.

Exercise 14.3.1. Suppose hunters kill 50 deer in a national forest every hunting season, while the deer by themselves have equal birth and death rates. If there are initially 500 deer in the forest, predict how many there will be in 5 years.

Exercise 14.3.2. Bacteria divide in two every hour, and half the current population dies (not including the new offspring). If the colony initially has 10 million bacteria, predict how many there will be in 24 hours.

Exercise 14.3.3. Each pair of rabbits produces 4 offspring every year, and the adults have a 0.5 annual death rate after reproduction. If initially there are 20 rabbits, how many do you predict there will be in 25 years?

Exercise 14.3.4. Consider the same rabbit population as in the previous exercise, but now a python that lives nearby eats exactly 1 rabbit a month. If initially there are 40 rabbits, how many do you predict there will be in 25 years?

Exercise 14.3.5. Five fish are added to an aquarium every month, while 90% of those present survive every month, and there is no reproduction. If initially there are 10 fish in the aquarium, find the solution and predict how many fish there will be in a year.

14.4 Numerical solutions in R

14.4.1 functions in R

Like most programming languages, R allows one to define and use structures called functions. Some are already written and loaded into the R distribution, for example, the function mean() we use to compute the mean of a vector variable, while others can be defined by users. Functions are discrete chunks of code that can be *called* from outside the function code to perform some task. The function receives inputs from the call and returns the result back. Here is the general structure of a function in R:

```
myfunction <- function(arg1, arg2, ... ){
   statements
   return(answer)
}
```

The first word "myfunction" is the name of the function, which will be used to call it. The word "function" is a keyword that R uses to identify the chunk of code as a function. Finally, in parentheses, is a comma-separated list of input arguments. Let's go ahead and make our first function. Copy the following code into a new script file, and run the script.

```
dot.product <- function(x, y) {
   # function to calculate the dot product of two
   # vectors
   if (length(x) == length(y)) {
      # check if the two vectors are the same length
      ans <- sum(x * y)
   } else {
      # if vectors are not the same length, give an error
      # message
      print("Vectors must be the same length!")
      ans <- NA
   }
   return(ans)
}
```

You should see the function name dot.product appear in the upper right window of R Studio—this means the function was created. However, it doesn't do anything by itself. To use the function, you need to call it by invoking its name. This you can do either from the command line, or from the script after the function has been read. Copy the following three lines into the script file, and then run it again:

```
vec1 <- runif(10)
vec2 <- 1:10
result <- dot.product(vec1, vec2)   # function call
```

Now the variable result contains the dot product of the two vectors vec1 and vec2. Note that the input argument in a function call doesn't have to have the same name it has in the definition of the function. Once the variables are *passed* inside the function, the function assigns vec1 and vec2 to x and y, respectively. When a function contains multiple input arguments, they are distinguished solely by the order in which they are passed to the function in the function call—for example, the first variable will be called x and the second will be called y, while the function is executing. So please pay attention to the order in which you write the input variables in a function call—if they are out of order, the function will not compute correctly, even if their names appear to be correct.

Programming exercises:

Write R functions to do the following tasks, and then test that they work by calling them and checking the results against the provided correct answers.

Exercise 14.4.1. Given input arguments a, k, and t, return the value of the exponential function $f(x) = ae^{kt}$. Self-check: given inputs of $a = 10$, $k = 2$, and $t = 10$, the function should return 4851651954.

Exercise 14.4.2. Given a, b, and c as input arguments, return the two roots of the quadratic function $f(x) = ax^2 + bx + c$ as a vector.

14.4. NUMERICAL SOLUTIONS IN R

Self-check: given inputs of $a = 1$, $b = 0$, and $c = -4$, the function should return the vector $(2, -2)$.

Exercise 14.4.3. Given input arguments a, k, and P, return the smallest integer value t at which $ae^{kt} > P$. Hint: k must be positive (exponential growth) and $P > a$, otherwise this doesn't work. There are multiple ways of solving this problem, the easiest is algebraic, but one can use a for loop or a while loop (which we haven't covered) to increase t until the function exceeds the value P. Self-check: for $a = 10$, $k = 2$, and $P = 10^9$, the answer is 10.

14.4.2 solving difference equations

Computers are naturally suited for precise, repetitive operations. One can use this ability to solve difference equations by iterating a given function to produce a sequence of values of the dependent variable x. We only need two things: to specify a computer function $f(x)$, which returns the value of the iterated map for any input value x, and the initial value X_0. Then it is a matter of repeating the operation of evaluating $f(X(t))$ and storing it as the next value $X(t + 1)$. Here is an outline of the calculation, but bear in mind that the mathematical notation in the outline is different from the programming syntax of R.

Calculation of a numerical solution of a difference equation:

1. Define the updating function $f(x)$ of the difference equation $x(t + 1) = f(x(t))$.

2. Set the initial value of the numerical solution $sol(1)$ and number of time steps N.

3. Execute a for loop for N steps.

4. Inside the for loop: set the next value of $sol(i + 1)$ to be $f(sol(i))$.

The resulting vector of values sol is called a *numerical solution* of the difference equation. The numerical method has two disadvantages compared to an analytic solution: first, the solution can only be obtained for a specific initial value and number of iterations, and second, any computer simulation inevitably introduces some errors, for instance, from round-off. In practice, however, most complex dynamical systems have to solved numerically, as analytic solutions are difficult or impossible to find.

Example. Let us solve the linear difference equation (the example of population doubling every time step):

$$N(t+1) = 2N(t)$$

The script below, defines a new function in R that corresponds to the updating function of a difference equation and uses it to solve the difference equation for 30 time steps. In the second part of the script the difference equation

$$N(t+1) = 0.8N(t)$$

is solved using the same method. Then the solutions are plotted for both equations in Figure 14.3 (script not shown). The solution of the first equation (variable `sol1`) grows exponentially, while the solution of the second equation (`sol2`) declines to zero.

```
updating.funk <- function(N) {
    ans <- 2 * N     # updating function f(N)
    return(ans)
}
numsteps <- 30   # set number of steps
sol1 <- rep(0, numsteps + 1)   # pre-allocate sol1
sol1[1] <- 100   # set initial value
time <- 0:numsteps   # define time vector
for (i in 1:numsteps) {
    # repeat for numsteps
    sol1[i + 1] <- updating.funk(sol1[i])
}
updating.funk <- function(N) {
    ans <- 0.8 * N   # updating function f(N)
    return(ans)
}
```

14.4. NUMERICAL SOLUTIONS IN R

```
sol2 <- rep(0, numsteps + 1)    # pre-allocate sol2
sol2[1] <- 100   # set initial value
time <- 0:numsteps    # define time vector
for (i in 1:numsteps) {
    # repeat for numsteps
    sol2[i + 1] <- updating.funk(sol2[i])
}
```

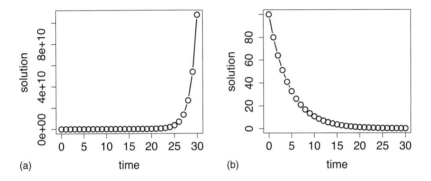

Figure 14.3: Numerical solutions of two linear difference equations: (a) $N(t+1) = 2N(t)$ and (b) $N(t+1) = 0.8N(t)$, both with initial value $N(0) = 100$.

Example. Let us now numerically solve a linear difference equation with a constant term, like ones solved analytically in section 14.2.

$$N(t+1) = -0.5N(t) + 15$$

We do not show the script for solving this difference equation, because it simply replicates the script above with a modified function updatingfunk. The unseen script solves this difference equation starting with two different initial values $N(0) = 20$ and $N(0) = 2$. The resulting solutions are shown in Figure 14.4, demonstrating that regardless of the initial value, the solutions approach the constant number $15/(1 - (-0.5)) = 10$, just as predicted in equation 14.3.

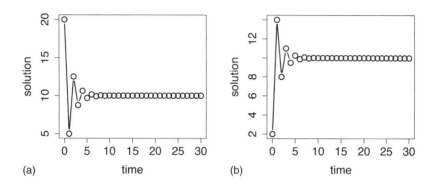

Figure 14.4: Solutions of the difference equation $N(t+1) = -0.5N + 15$ starting with different initial values: (a) $N(0) = 20$ and (b) $N(0) = 2$.

Programming exercises:

Write an R function to do each of the following programming tasks.

Exercise 14.4.4. Take the input value, multiply it by 1.03, and return the result. If the initial value is 5, the new value should be 5.15.

Exercise 14.4.5. Take the input value, multiply it by 1.03 as many times as given by the other input argument, and return the result. Hint: Use a for loop to do this, and check that the second value is a positive integer before using it. Starting with the initial value is 5 and 100 iterations, the function should return the value 96.093.

Exercise 14.4.6. Modify the previous function to save all values into a vector, and return this vector. Plotting the vector should result in an exponential graph.

14.5 Computational project

In this project you will calculate numerical solutions of difference equations. Follow the outline in subsection 14.4.2 to create a code that will iteratively update the variable for a specified number of

14.5. COMPUTATIONAL PROJECT

steps, store it as a vector, and plot it as function of time. Hint: Your code will be nearly identical for all the tasks, so once you get the first script to work properly, copy and paste it, and modify the line that updates the variable (the updating function).

Tasks

For the following difference equations: (a) calculate the numerical solution for 100 steps staring with the given initial values; (b) plot the solution over time; (c) describe how the solution behaves in the long term, and determine whether it depends on the initial value.

1. Initial values $X(0) = 20$; $X(0) = 0$:
$$X(t+1) = 45 - 0.8X(t)$$

2. Initial values $X(0) = 3$; $X(0) = 1$:
$$X(t+1) = \frac{2}{X(t)+1}$$

3. Initial values $X(0) = 2$; $X(0) = 9$:
$$X(t+1) = 0.2X(t)(10 - X(t))$$

4. Initial values $X(0) = 2$; $X(0) = 9$:
$$X(t+1) = 0.4X(t)(10 - X(t))$$

5. Initial values $X(0) = 2$; $X(0) = 9$:
$$X(t+1) = X(t)(10 - X(t))$$

Chapter 15

Linear ordinary differential equations

> *He felt a restless, vague ambition,*
> *A craving for a change of air*
> *(A most unfortunate condition,*
> *A cross not many choose to bear.)*
> —Alexander Pushkin, *Eugene Onegin*

In Chapter 14 we considered discrete time models, in which time is counted in integers. This worked well to describe processes that happen in periodic cycles, like cell division or heart pumping. Many biological systems do not work this way. Change can happen continuously, that is, at any moment in time. For instance, the concentration of a biological molecule in the cell changes gradually, as does the voltage across the cell membrane in a neuron.

There are at least two good reasons to use differential equations for many applications. First, they are often more realistic than discrete time models, because some events happen frequently and non-periodically. The second reason is mathematical: it turns out that dynamical systems with continuous time, described by differential equations, are better behaved than difference equations. This has to do with the essential "jumpiness" of difference equations. Even for simple nonlinear equations, the value of the variable after one time step can be far removed from its previous value. This can lead to highly complicated solutions, as we saw in some of the numerical solutions in Chapter 14. That kind of erratic behavior is impossible in one-variable differential equations (provided they have continuous defining functions). In this chapter you will learn to do the following:

1. build differential equations based on stated assumptions,

2. find analytic solutions of linear differential equations, and

3. compute numerical solutions of differential equations using the Forward Euler method.

15.1 From discrete time to smooth change

15.1.1 bacteria that divide at arbitrary times

In this chapter we investigate *continuous time dynamical systems*, for which it does not make sense to break time up into equal intervals. Instead of equations describing the increments in the dependent variable from one time step to the next, we will use equations with an instantaneous rate of change (derivative) of the variable. Let us see the connection between the discrete and continuous dynamic models by reducing the step size of the bacteria-division population model.

First, suppose that instead of dividing every hour, the population of bacteria divides every half-hour, but only half of the population does so. That half is chosen randomly, so we don't have to keep track of whether each bacterium divided the last time around. Therefore, each half-hour, exactly half the population is added to the current population:

$$N(t+0.5) = N(t) + 0.5N(t) - 1.5N(t)$$

The solution for this model can be figured out from the linear difference equation solution derived in section 14.2. Every half-hour, the population is multiplied by 1.5, so we can write

$$N(t) = 1.5^{2t} N(0) = (1.5^2)^t N(0)$$

15.1. FROM DISCRETE TIME TO SMOOTH CHANGE

Compare this solution with the one for the every-hour model, $N(t) = 2^t N(0)$ by plugging in a few numbers for t. The half-hour model grows faster, because it has the base of 2.25 instead of 2.

Now, suppose that the bacteria can divide four times an hour, but only a quarter of the population reproduces at any given time. The model can be written similarly:

$$N(t + 0.25) = N(t) + 0.25N(t) = 1.25N(t)$$

The solution for this model is once again exponential, with the difference that each hour contains 4 division events:

$$N(t) = 1.25^{4t} N(0) = (1.25^4)^t N(0)$$

This solution has the exponential base is 1.25^4, which is larger than 1.5^2. So what happens when we take this further?

Suppose the bacteria divide m times an hour, with time step $1/m$. Then extending our models, we can write down the model and the solution:

$$N(t + 1/m) = N(t) + (1/m)N(t) = (1 + 1/m)N(t)$$

$$N(t) = (1 + 1/m)^{mt} N(0) = [(1 + 1/m)^m]^t N(0)$$

Now we can do what mathematicians enjoy the most: take things to the limit. What if m were 100? A million? A gazillion? Let us rewrite the model equation:

$$N(t + 1/m) - N(t) = 1/m N(t) \Rightarrow \frac{N(t + 1/m) - N(t)}{1/m} = N(t)$$

The expression on the left is known as Newton's quotient, which you encounter in the definition of a derivative. It measures the rate of change of the population N from some time t to the next time step $t + 1/m$. If m is increased to make the time step smaller, this makes both the numerator and the denominator smaller, and the quotient approaches the *instantaneous rate of change* of $N(t)$. So, if bacteria divide at any point in time, with the average rate of 1 per hour, the model becomes a differential equation:

$$\frac{dN}{dt} = N(t)$$

We can do a similar procedure to the formula of the solution of the model. The dependence on m is all on the left-hand side, in the expression $(1 + 1/m)^m$, which is the base of the exponential function. What happens to this number as m becomes larger? Does it increase without bound? You can investigate this numerically by plugging in progressively larger numbers m. You will see that the number approaches a specific value, about 2.71828. This is the special constant e, called the base of the natural logarithm. So, if bacteria divide at any point in time, with the average rate of 1 per hour, the solution of the model becomes:

$$N(t) = e^t N(0)$$

Math exercises:

These exercises illustrate the effect of changing the time step on a difference equation model, and prepare the ground for differential equation models.

Exercise 15.1.1. Go back to the model of rabbits where each pair produces 4 offspring (assume all rabbits are breeding), and then 0.5 of the existing (adult) population dies, and both births and deaths happen every year. Write down the difference equation using the increment form: $N(t+1) - N(t) = $ some expression. Find the solution of the model (with a generic initial value of rabbits)

Exercise 15.1.2. Now suppose that the rabbits breed and die at the same rate, but that they reproduce and die every month, so each month, 1/12 of the yearly births and deaths occur. Write down the difference equation in the increment form $N(t+1) - N(t) = $ some expression, with time measured in months. Find the solution of the model, change the units to express the time in years, and compare this solution with the one for the model in the first exercise.

Exercise 15.1.3. Keep the same rates of births and death, but now allow them to reproduce and die every day, so every day 1/365 of the yearly births and deaths occur. Write down the difference equation in the increment form $N(t+1) - N(t) = $ some expression,

15.1. FROM DISCRETE TIME TO SMOOTH CHANGE 329

with time measured in days. Find the solution of the model, change the units to express the time in years, and compare this solution with the one for the model in first exercise.

Exercise 15.1.4. Now try extending this trend to the mathematical limit. What happens if the rabbits can reproduce and die at any time (make the time step extremely small)? What does the solution of that model approach as the time step shrinks to zero? (Do some numerical experimentation to figure it out).

15.1.2 growth proportional to population size

We will now build some of the most common differential equations models. First up is a simple population growth model with a constant growth rate. Suppose that in a population, each individual reproduces with the average reproductive rate r. This is reflected in the following differential equation:

$$\frac{dx}{dt} = \dot{x} = rx \tag{15.1}$$

This expression states that the rate of change of x, which we take to be population size, is proportional to x with multiplicative constant r. We sometimes use the notation $\dot{x}$ for the time derivative of x (which was invented by Newton) for aesthetic reasons.

First, we apply dimensional analysis to this model. The units of the derivative are population per time, as can be deduced from the Newton's quotient definition. Thus, the units in the equation have the following relationship:

$$\frac{[population]}{[time]} = [r][population] = \frac{1}{[time]}[population]$$

This shows that as in the discrete time models, the dimension of the population growth rate r is inverse time, or frequency. The difference between the differential and discrete time population models lies in the time scope of the rate. In the case of the difference equation, r is the rate of change per one time step of the model. In the differential equation, r is the *instantaneous rate of population* growth. It is less intuitive than the growth rate per single reproductive cycle, just like the slope of a curve is less intuitive than the slope of a line. If population growth happens continuously, the growth rate of r individuals per year does not mean that if we start with one individual, there will be r after one year. To make quantitative predictions, we need to find the solution of the equation, as is done in the next section.

15.1.3 chemical kinetics

Reactions between molecules in cells occur continuously, driven by molecular collisions and physical forces. To model this complex behavior, it is generally assumed that reactions occur with a particular speed, known as the *kinetic rate constant*. A simple reaction of conversion from one type of molecule (A) to another (B) can be written as

$$A \xrightarrow{k} B$$

In this equation the parameter k is the kinetic rate rate constant, describing the speed of conversion of A into B per concentration of A.

Chemists and biochemists use differential equations to describe the change in molecular concentration during a reaction. These equations are known as the *laws of mass action*. For the reaction above, the concentration of molecule A decreases continuously proportionally to itself, and the concentration of molecule B increases continuously proportionally to the concentration of A. This is expressed by the following two differential equations:

$$\dot{A} = -kA \qquad (15.2)$$
$$\dot{B} = kA \qquad (15.3)$$

Several conclusions are apparent by inspection of these equations. First, the dynamics depend only on the concentration of A, so keeping track of the concentration of B is superfluous. The second observation reinforces the first: the sum of the concentrations of A and B is constant. This is mathematically demonstrated by adding the two equations together to obtain the following:

$$\dot{A} + \dot{B} = -kA + kA = 0$$

15.2. SOLUTIONS OF ODES

One of the basic properties of the derivative is that the sum of derivatives is the same as the derivative of the sum:

$$\dot{A} + \dot{B} = \frac{d(A+B)}{dt} = 0$$

This means that the sum of the concentrations of A and B is a constant. This is a mathematical expression of the law of conservation in chemistry: molecules can change from one type to another, but they cannot appear or disappear in other ways. In this case, a single molecule of A becomes a single molecule of B, so it follows that the sum of the two has to remain the same. If the reaction were instead two molecules of A converting to a molecule of B, then the conserved quantity is $2A + B$. The concept of conserved quantity is very useful for the analysis of differential equations, espically for simplifying nonlinear differential equations.

15.2 Solutions of ordinary differential equations

In this section we investigate how to write down analytic solutions for ordinary differential equations (ODEs). A differential equation is an equation that contains derivatives of the dependent variable (which we usually call x). For the time being, we restrict ourselves to ODEs with the highest derivative being of first order. In general, we can write all such ODE as

$$\frac{dx}{dt} = \dot{x} = f(x, t)$$

Note that the function may depend on both the dependent variable x and the independent variable t. Let us first define some terminology for ODEs:

Definition 15.1. The *order* of an ODE is the highest order of the derivative of the dependent variable x.

For example, $\dot{x} = rx$ is a first-order ODE, while $\ddot{x} = -mx$ is a second-order ODE (double dot stands for second derivative).

Definition 15.2. An ODE is *autonomous* if the function f depends only on the dependent variable x and not on t.

For example, $\dot{x} = 5x - 4$ is an autonomous equation, while $\dot{x} = 5t$ is not. An autonomous ODE is also said to have *constant coefficients* (e.g., 5 and -4 in the first equation above).

Definition 15.3. An ODE is *homogeneous* if every term involves either the dependent variable x or its derivative.

For instance, $\dot{x} = x^2 + \sin(x)$ is homogeneous, while $\dot{x} = -x + 5t$ is not.

Most simple biological models discussed in Chapters 15 and 16 are autonomous, homogeneous ODEs. However, inhomogeneous equations are important in many applications, and we will encounter them at the end of the present section.

15.2.1 separate-and-integrate method

The *solution of a differential equation* is a function of the independent variable that satisfies the equation for a range of values of the independent variable. In contrast to algebraic equations, we cannot simply isolate x on one side of the equal sign and find the solutions as one, or a few numbers. Instead, solving ordinary differential equations is tricky, and no general strategy for solving an arbitrary ODE exists. Moreover, a solution for an ODE is not guaranteed to exist at all, or not for all values of t. We will discuss some of the difficulties later, but let us start with equations that we can solve.

The most obvious strategy for solving an ODE is integration. Since a differential equation contains derivatives, integrating it can remove the derivative. In the case of the general first-order equation, we can integrate both sides to obtain

$$\int \frac{dx}{dt} dt = \int f(x,t) dt \Rightarrow x + C = \int f(x,t) dt$$

The constant of integration C appears as in the standard antiderivative definition. It can be specified by an initial condition for the solution $x(t)$.

15.2. SOLUTIONS OF ODES

The simplest method of analytical solution of first-order ODEs, which I call the *separate-and-integrate* method, consists of the following steps.

1. Use algebra to place the dependent and independent variables on different sides of the equations, including the differentials (e.g., dx and dt).

2. Integrate both sides with respect to the different variables, and don't forget the integration constant.

3. Solve for the dependent variable (e.g., x).

4. Plug in $t = 0$, and use the initial value $x(0)$ to solve for the integration constant.

Example. Consider the simple differential equation: $\dot{x} = a$, where $\dot{x}$ stands for the time derivative of the dependent variable x, and a is a constant. It can be solved by integration:

$$\int \frac{dx}{dt} dt = \int a dt \Rightarrow x(t) + C = at$$

This solution contains an undetermined integration constant; if an initial condition is specified, we can determine the solution. Generally speaking, if the initial condition is $x(0) = x_0$, we need to solve an algebraic equation to determine C: $x_0 = a \times 0 - C$, which results in $C = -x_0$. The complete solution is then $x(t) = at + x_0$. To make the example more specific, if $a = 5$ and the initial condition is $x(0) = -3$, the solution is $x(t) = 5t - 3$.

Example. Let us solve the linear population growth model given by equation 15.1: $\dot{x} = rx$. The equation can be solved by first dividing both sides by x and then integrating:

$$\int \frac{1}{x} \frac{dx}{dt} dt = \int \frac{dx}{x} = \int r dt \Longrightarrow \log|x| = rt + C \Longrightarrow x$$
$$= e^{rt+C} = Ae^{rt}$$

We used basic algebra to solve for x, exponentiating both sides to get rid of the logarithm on the left side. As a result, the additive constant C gave rise to the multiplicative constant $A = e^C$. Once again, the solution contains a constant, which can be determined by specifying an initial condition $x(0) = x_0$. In this case, the relationship is quite straightforward: $x(0) = Ae^0 = A$. Thus, the complete solution for equation 15.1 is:

$$x(t) = x_0 e^{rt}$$

As in the case of the discrete-time models, population growth with a constant birth rate has exponential form. Once again, please pause and consider this fact, because the exponential solution of linear equations is one of the most basic and powerful tools in applied mathematics. It immediately allows us to classify the behavior of linear ODE into three categories:

- $r > 0$: $x(t)$ grows without bound
- $r < 0$: $x(t)$ decays to 0
- $r = 0$: $x(t)$ remains constant at the initial value

The rate r being positive reflects the dominance of birth rate over death rate in the population, leading to unlimited population growth. If the death rate is greater, the population will decline and die out. If the two are exactly matched, the population size will remain unchanged.

Example. The solution for the biochemical kinetic model given by equation 15.2 is identical to that for the population growth model except for the sign: $A(t) = A_0 e^{-kt}$. When the reaction rate k is positive, as it is in chemistry, the concentration of A decays to 0 over time. This should be obvious from our model, since there is no back reaction, and the only chemical process is conversion of A into B. The concentration of B can be found by using the fact that the total concentration of molecules in the model is conserved. Let us call it C. Then $B(t) = C - A(t) = C - A_0 e^{-kt}$. The concentration of B increases to the asymptotic limit of C, meaning that all molecules of A have been converted to B.

15.2.2 solution of inhomogeneous ODEs

The two models we have analyzed above are homogeneous, because every term contains the dependent variable, and linear, because the relationship with the dependent variable is proportional. Inhomogeneous ODEs can be solved on paper using the same separate-and-integrate method, modified slightly to handle the constant term. Here are the steps to solve the generic linear ODE with a constant term $\dot{x} = ax + b$:

1. Separate the dependent and independent variables on different sides of the equations by dividing both sides by the right-hand side $ax + b$ and multiplying both sides by the differential dt.

2. Integrate both sides with respect to the different variables, don't forget the integration constant!

3. Solve for the dependent variable (e.g., x).

4. Plug in $t = 0$, and use the initial value $x(0)$ to solve for the integration constant.

Example. Let us solve the following ODE model using the separate-and-integrate method with the given initial value:

$$\frac{dx}{dt} = 4x - 100; \quad x(0) = 32$$

1. Separate the dependent and independent variables:

$$\frac{dx}{4x - 100} = dt$$

2. Integrate both sides:

$$\int \frac{dx}{4x - 100} = \int dt \Rightarrow \frac{1}{4} \int \frac{du}{u} = \frac{1}{4} \ln|4x - 100| = t + C$$

The integration uses the substitution of the new variable $u = 4x - 100$, with the concurrent substitution of $dx = du/4$.

3. Solve for the dependent variable:

$$\ln|4x - 100| = 4t + C \Rightarrow 4x - 100 = e^{4t}B \Rightarrow x = 25 + Be^{4t}$$

Here the first step is to multiply both sides by 4, and the second is to exponentiate both sides, removing the natural log from the left-hand side, and finally to use simple algebra to solve for x as a function of t.

4. Solve for the integration constant:

$$x(0) = 25 + B = 32 \Rightarrow B = 7$$

Here the exponential disappeared because $e^0 = 1$.

Therefore, the complete solution of the ODE with the given initial value is

$$x(t) = 25 + 5e^{4t}$$

At this point, you might have noticed something about solutions of linear ODEs: they always involve an exponential term, with time in the exponent. Knowing this, it is possible to bypass the whole process of separate-and-integrate by using the following shortcut.

Important fact: Any linear ODE of the form $\dot{x} = ax + b$ has an analytic solution of the form:

$$x(t) = Ce^{at} + D$$

This can be tested by plugging the solution back into the ODE to see whether it satisfies the equation. First, take the derivative of the solution to get the left-hand side of the ODE: $\frac{dx}{dt} = Cae^{at}$; the plug in $x(t)$ into the right-hand side of the ODE: $aCe^{at} + aD + b$. Equating the two sides, we get $Cae^{at} = aCe^{at} + aD + b$, which is satisfied if $aD + b = 0$, which means $D = -b/a$. This is consistent with the example above: the additive constant in the solution was 0.8, which is $-b/a = -(4)/5 = 0.8$.

In short, if you want to solve a linear ODE $\dot{x} = ax + b$, you can bypass the separate-and-integrate process, because the general solution always has the form:

$$x(t) = Ce^{at} - \frac{b}{a} \tag{15.4}$$

15.2. SOLUTIONS OF ODES

The unknown constant C can be determined from a given initial value. So the upshot is that all linear ODEs have solutions that are exponential in time with the exponential constant coming from the slope constant a in the ODE. The dynamics of the solution are determined by the sign of the constant a: if $a > 0$, the solution grows (or declines) without bound; and if $a < 0$, the solution approaches an asymptote at $-b/a$ (from above or below, depending on the initial value). Go back and read section 2.2 for a review of exponential functions if this is not clear.

Math exercises:

Solve the following linear ODEs and use the specified initial values to determine the integration constant. Describe how the solution behaves over a long time (e.g., grows without bound, goes to zero.). Plug the solution back into the ODE to check that it satisfies the equation.

Exercise 15.2.1.
$$\frac{dx}{dt} = -5; \ x(0) = -7$$

Exercise 15.2.2.
$$\frac{dx}{dt} = \cos(4t); \ x(0) = -7$$

Exercise 15.2.3.
$$\frac{dx}{dt} = -2x; \ x(0) = 100$$

Exercise 15.2.4.
$$\frac{dx}{dt} = \frac{1}{3}x; \ x(0) = -30$$

Exercise 15.2.5.
$$\frac{dx}{dt} = 2(1-x); \ x(0) = 2$$

Exercise 15.2.6.
$$\frac{dx}{dt} = 5 - 0.5x; \ x(0) = 20$$

15.2.3 Forward Euler method

Analytic solutions are very useful for a modeler, because they allow prediction of the variable of interest at any time in the future. However, for many differential equations, they are not easy to find; for many others, they simply cannot be written down. Instead, one can use a numerical approach, which does not require an exact formula for the solution. The idea is to start at a given initial value (e.g., $x(0)$) and use the derivative from the ODE (e.g., dx/dt) as the rate of change of the solution (e.g., $x(t)$) to calculate the change or increment for the solution over a time step. Essentially, this means replacing the continuous change of the derivative with a discrete time step, thus converting the differential equation into a difference equation and then solving it. The solution of the difference equation is not the same as the solution of the ODE, so *numerical solutions of ODEs are always approximate*. I will use the notation $y(t)$ to denote the numerical solution to distinguish it from the exact solution $x(t)$. The fundamental difference between them is that $y(t)$ is not a formula that can be evaluated at any point in time, but instead is a sequence of numbers calculated every time step, which hopefully are close to the exact solution $x(t)$.

Let us introduce all the players. First, we need to pick the time step Δt, which is the length of time between successive values of y. In the difference equation notation, one can use y_i to mean $y(i\Delta t)$, the value of the numerical solution after i time steps. Then we need to calculate the derivative, or the rate of change at a particular point in time. For any first-order ODE of the form

$$\frac{dx}{dt} = \dot{x} = f(x,t)$$

the rate of change depends (potentially) on the values of x and t. This rate of change based on the numerical solution after i time steps is $f(y(i\Delta t), i\Delta t) = f(y_i, t_i)$. Finally, to calculate the change of the dependent variable, we need to multiply the rate of change by the time step. This should make sense in a practical context: if you drive for two hours (time step) at 60 miles per hour (rate of change), the total distance (increment) is $2 \times 60 = 120$ miles. By

15.2. SOLUTIONS OF ODES

the same token, we can write down how to calculate the next value of the numerical solution y_{i+1} based on the previous one:

$$y_{i+1} = y_i + \Delta t f(y_i, t_i) \qquad (15.5)$$

This method of computing a numerical solution of an ODE is called the *Forward Euler method*, after the famous mathematician who first came up with it. It is called a forward method, because it uses the value of the dependent variable and its derivative at time step i to predict the value at the next time step $i+1$. The method is iterative, so it needs to be repeated to calculate a set of values of the approximate solution $y(t)$. Here are a couple of simple examples of computing numerical solutions using Forward Euler (FE).

Example. Numerically solve the ODE $\dot{x} = -0.1$ using the Forward Euler method. The defining function in the formulation given above is $f(x,t) = -0.1$. We can calculate the numerical solution for a couple of steps and compare the values with the exact solution, since we now know that the latter is $x(t) = x_0 - 0.1t$. Let us pick the time step $\Delta t = 0.2$ and begin with the initial value $x(0) = 1$. Here are the first three steps:

$$y(0.2) = y(0) + \Delta t f(y(0)) = 1 + 0.2 \times (-0.1) = 0.98$$
$$y(0.4) = y(0.2) + \Delta t f(y(0.2)) = 0.98 + 0.2 \times (-0.1) = 0.96$$
$$y(0.6) = y(0.4) + \Delta t f(y(0.4)) = 0.96 + 0.2 \times (-0.1) = 0.94$$

Since the rate of change in this ODE is constant, the solution declines by the same amount every time step. In this case, the numerical solution is actually exact—it perfectly matches the analytic solution. Table 15.1 (left) shows the numerical solution for three time steps along with the exact solution.

Example. Numerically solve the ODE $\dot{x} = -0.1x$ using the Forward Euler method. The defining function is $f(x,t) = -0.1x$. We can calculate the numerical solution for a couple of steps and compare the values with the exact solution, since we now know that the latter is $x(t) = x_0 e^{-0.1t}$. Let us pick the time step $\Delta t = 0.2$ and begin with the initial value $x(0) = 100$. Here are the first three steps:

$$y(0.2) = y(0) + \Delta t f(y(0)) = 100 + 0.2 \times (-0.1 \times 100) = 98$$
$$y(0.4) = y(0.2) + \Delta t f(y(0.2)) = 98 + 0.2 \times (-0.1 \times 98) = 96.04$$
$$y(0.6) = y(0.4) + \Delta t f(y(0.4)) = 96.04 + 0.2 \times (-0.1 \times 96.04) \approx 94.12$$

In this case, the derivative is not constant, and the numerical solution is not exact, which is demonstrated in Table 15.1 (right). The error in the numerical solution grows with time, which may be problematic. We will further investigate how to implement the computation of numerical solutions using R in section 15.3.

t	x	y	error	t	x	y	error
0	1	1	0	0	100	100	0
0.2	0.98	0.98	0	0.2	98.02	98	0.02
0.4	0.96	0.96	0	0.4	96.08	96.04	0.04
0.6	0.94	0.94	0	0.6	94.18	94.12	0.06

Table 15.1: Numerical solutions y of two ODEs: $\dot{x} = -0.1$ (left) and $\dot{x} = -0.1x$ (right) using Forward Euler calculated for three steps of size $\Delta t = 0.2$ as well as the exact solution x (both rounded to two digits after the decimal) and the error of the numerical solution.

Math exercises:

Use the Forward Euler method to solve the following differential equations with time step $\Delta t = 0.1$ for two steps to compute $y(0.2)$ (the value of the numerical solution at $t = 0.2$).

Exercise 15.2.7.
$$\frac{dx}{dt} = -5; \ x(0) = -7$$

Exercise 15.2.8.
$$\frac{dx}{dt} = \cos(4t); \ x(0) = -7$$

Exercise 15.2.9.
$$\frac{dx}{dt} = -2x; \ x(0) = 100$$

Exercise 15.2.10.
$$\frac{dx}{dt} = \frac{1}{3}x; \ x(0) = -30$$

Exercise 15.2.11.
$$\frac{dx}{dt} = 2(1-x); \ x(0) = 2$$

Exercise 15.2.12.
$$\frac{dx}{dt} = 5 - 0.5x; \ x(0) = 20$$

15.3 Numerical solutions of ODEs

15.3.1 implementation of Forward Euler

In practice, the most common approach to finding solutions for differential equations is using a computer to calculate a numerical solution, for example, using the Forward Euler method. This means using a computer program to construct a sequence of values of the dependent variable that approximate the true solution. Below is an outline of an algorithm that can be translated into a programming language, like R, to solve ODEs:

Outline of Forward Euler method:

1. Define the function $f(x,t)$ that returns the value of the derivative (dx/dt).

2. Set the time step Δt and the total time T.

3. Set the number of time steps N to be $T/\Delta t$.

4. Pre-allocate a vector of numeric solutions y of length $N = 1$ and set $y[1]$ to the initial value.

5. Create a vector of time values t from 0 to N with time step Δt.

6. Create a for loop for N steps.

7. Inside the for loop: set the next value of $y[i+1]$ to be $y[i] + \Delta t \times f(y[i], t[i])$.

Like any algorithm, you need to be clear about its inputs and outputs. In this case, the inputs are the defining function $f(x,t)$, the initial value, the time step, and the total time. The output is the solution vector y, which contains a sequence of values that approximate the solution of the ODE, along with the vector of time values spaced by the time step. Below is an example of an implementation of the Forward Euler method in R. Plots of numerical solutions produced with this script for two different linear ODEs are shown in Figure 15.1.

```
def.funk <- function(a, x) {
    ans <- a * x
    return(ans)
}
Tmax <- 10    # length of time
dt <- 0.2     # time step
numsteps <- Tmax/dt    # number of steps
y <- rep(0, numsteps + 1)    # pre-allocate y
time <- seq(0, Tmax, dt)    # set time vector
a <- 0.5    # set the parameter value
y[1] <- 100    # initial value
for (i in 1:numsteps) {
    # Forward Euler
    y[i + 1] <- y[i] + dt * def.funk(a, y[i])
}
```

Notice that this script is similar to the one for numerical solution of a difference equation section in 14.4, the major difference being the presence of a time step, whereas in difference equations the time step is aways 1. There is one more important point for the

15.3. NUMERICAL SOLUTIONS OF ODES

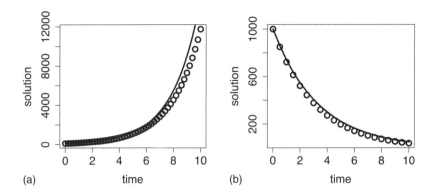

Figure 15.1: Numerical solutions using Forward Euler (circles) and the analytic solutions (continuous curves) of two ODEs: (a) $dx/dt = 0.5x$ with $x(0) = 100$ and $\Delta t = 0.2$; (b) $dx/dt = -0.3x$ with $x(0) = 1000$ and $\Delta t = 0.5$.

implementation: usually one needs to solve the ODE for a particular length of time T with a specified time step Δt. This dictates that the required number of iterations be $T/\Delta t$; in other words, for a given time period the number of time steps is inversely proportional to the time step.

15.3.2 error in Forward Euler solutions

One of the main concerns of numerical analysis is to minimize the *error*: the difference between the true solution and the numerical solution. There are at least two distinct sources of error in numerical solutions: (1) *roundoff error* and (2) *truncation error*. Roundoff error is caused by computers representing real numbers by a finite string of bits on a computer using what is known as a *floating point* representation. In many programming languages, variables storing real numbers can be declared as single or double precision, which typically support 24 and 53 significant binary digits, respectively. Any arithmetic operation involving floating point numbers is only approximate, with an error that depends on the way the numbers are stored in memory. Truncation error is caused by approximations inherent in numerical algorithms. The most common class of numerical approximations for ODEs is known as *finite difference* methods, and Forward Euler is a simple representative of that class. As the name suggests, these methods use difference equations to approximate a differential equation. There is inevitably a truncation error in such methods, because they use a more or less clever scheme to approximate the instantaneous rate of change in an ODE with a discrete time expression.

A numerical modeler has different controls over the roundoff error and truncation error. The first can be minimized by using more memory to store the numbers (e.g., by using double precision format for the variables). Further, there are techniques for minimizing the so-called loss of significance that occurs in certain arithmetic operations, like subtraction of two numbers of similar value. We will leave these considerations to numerical analysts (Press et al. 2007); for the most part, roundoff error is not a significant issue on modern computers. Truncation error, however, is much more within our control, because it depends on the choice of the numerical algorithm. One can decrease the error in the case of finite difference methods by choosing smaller time steps, or by choosing an algorithm with a higher *order of accuracy*.

Returning specifically to the Forward Euler method, it is called a *first-order method*, because the total error in the solution (after some number of time steps) depends linearly on the time step Δt. One can show this by using the Taylor expansion of the solution $y(t)$ to derive the Forward Euler method, with $\tau(\Delta t)$ representing the truncation error after one time step:

$$y(t + \Delta t) = y(t) + \Delta t \frac{dy(t)}{dt} + \tau(\Delta t)$$

As you might have learned in calculus, the error remaining after the linear term in the Taylor series is proportional to the the square of the small deviation Δt. This only describes the error after one time step, but since the errors accumulate every time step, the total error after N time steps accumulates as $N\tau(\Delta t)$. As we saw in the implementation above, for a given length of time, N is inversely proportional to Δt. Therefore, the total error is proportional to the Δt and so FE is a first-order method.

15.3. NUMERICAL SOLUTIONS OF ODES

The exercise above shows that new errors in FE method accumulate in proportion with the time step. What happens to these errors over time? Do they grow or dissipate with more iterations? This is known as the *stability* of a numerical method, and unlike the above question about the order of accuracy, the answer depends on the particular ODE that one needs to solve. Here is an example of error analysis for a linear ODE.

Example. To numerically solve the equation $\dot{x} = ax$, we substitute the function ax for the function $f(x,t)$, and obtain the FE approximation for this particular ODE:

$$y_{i+1} = y_i + \Delta t a y_i = (1 + a\Delta t)y_i$$

The big question is what happens to the truncation error. Does it grow or decay? To investigate this question, let us denote the error at time t_i, that is, the difference between the true solution $x(t_i)$ and the approximate solution $y(t_i)$, by ϵ_i. It follows that $y_i = x_i + \epsilon_i$. Then we can write the following difference equations involving the error:

$$y_{i+1} = x_{i+1} + \epsilon_{i+1} = (x_i + \epsilon_i)(1 + a\Delta t)$$
$$= x_i(1 + a\Delta t) + \epsilon_i(1 + a\Delta t)$$

Set aside the terms in the equation that involve x (since it is just the equation for Forward Euler). The remaining difference equation for ϵ describes the change in the error:

$$\epsilon_{i+1} = \epsilon_i(1 + a\Delta t)$$

This states that the error in this numerical solution is repeatedly multiplied by the constant $(1 + a\Delta t)$. As we saw in section 14.2, this linear difference equation has an exponential solution $\epsilon_n = (1 + a\Delta t)^n \epsilon_0$, which decays to 0 if $|1 + a\Delta t| < 1$ or grows without bound if $|1 + a\Delta t| > 1$. The first inequality is called the *stability condition* for the FE scheme, since it guarantees that the old errors decay over time. Since $\Delta t > 0$, the only way that the left-hand side can be less than 1 is if $a < 0$. Therefore, the condition for stability of the FE method for a linear ODE is

$$|1 + a\Delta t| < 1 \Rightarrow \Delta t < -2/a$$

Thus, if $a > 0$, the errors will eventually overwhelm the solution. If $a < 0$, if the time step is small enough (less than $-2/a$), then FE is stable. Generally speaking, however, Forward Euler is about the worst method to use for practical numerical solutions of ODEs, due to its low accuracy and to its lack of stability under certain conditions.

Programming exercises:

Complete these exercises by implementing Forward Euler method in R to solve the following ODE:

$$\frac{dx}{dt} = 0.2x$$

Exercise 15.3.1. Calculate the numerical solution of the ODE for one time step using Forward Euler, for time step $dt = 0.1$, starting with initial value $x(0) = 5$. Answer: 5.1.

Exercise 15.3.2. Calculate the error in the numerical solution above by subtracting it from the exact (analytic) solution $x(t) = e^{0.2t}x(0)$, with the same initial value. Answer: ≈ 0.001.

Exercise 15.3.3. Write a script to solve the ODE using the Forward Euler method based on the outline above. Set the time step $dt = 0.1$, and report the solution after 100 time steps (total time $T = 10$). Answer: 36.22323.

Exercise 15.3.4. Change the time step to be $dt = 0.01$, and report the solution after 1000 time steps (total time $T = 10$). Answer: 36.87156.

Exercise 15.3.5. Compute the errors of the numerical solutions at $t = 10$ with different time steps by subtracting them from the values obtained from the analytic solution at the same time. Answers: for $dt = 0.1$, the error is 0.722; for $dt = 0.01$, the error is 0.0737.

15.4 Applications of linear ODE models

15.4.1 model of pharmacokinetics

Describing and predicting the dynamics of drug concentrations in the body is the goal of *pharmacokinetics*. Any drug that humans take goes through several stages: first it is administered (put into the body), then absorbed, metabolized (transformed), and excreted (removed from the body) (Rosenbaum 2011). Almost any drug has a dose at which it has a toxic effect, and most can kill a human if the dose is high enough. Drugs used for medical purposes have a *therapeutic range*, which lies between the lowest possible concentration (usually measured in the blood plasma) that achieves the therapeutic effect and the concentration that is toxic. One of the basic questions that medical practitioners need to know is how much and how frequently to administer a drug to maintain drug concentration in the therapeutic range.

The concentration of a drug is a dynamic variable that depends on the rates of several processes, most directly on the rate of administration and the rate of metabolism. Drugs can be *administered* through various means (e.g., orally or intravenously) which influences their rate of absorption and thus how the concentration increases. Once in the blood plasma, drugs are metabolized primarily by enzymes in the liver, converting drug molecules into compounds that can be excreted through the kidneys or the large intestine. The process of *metabolism* proceeds at a rate that depends on both the concentration of the drug and the enzyme that catalyzes the reaction. For some drugs the metabolic rate may be *constant*, or independent of the drug concentration, since the enzymes are already working at full capacity and can't turn over any more reactions (e.g., alcohol is metabolized at a constant rate of about 1 drink per hour for most humans). Figure 15.2a shows the time plots of the blood alcohol concentration for four men who ingested different amounts of alcohol; the curves are essentially linear with the same slope after the peak. For other drugs, if the plasma concentration is low enough, the enzymes are not occupied all the time, and increasing the drug concentration leads to an increase in the rate of metabolism. One can see this behavior in the metabolism of the antidepressant drug bupropion in Figure 15.2b, where the

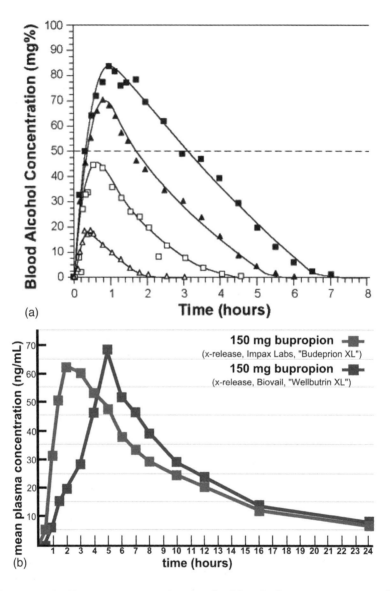

Figure 15.2: Drug concentration in the blood plasma over time: (a) Blood alcohol content after ingesting different numbers of drinks, from four in the top curve to one in the bottom (figure from the National Institute on Alcohol Abuse and Alcoholism in public domain); (b) Blood concentration of bupropion for two different drugs in clinical trials (image by CMBJ based on FDA data under CC-BY 3.0 via Wikimedia Commons).

15.4. APPLICATIONS OF LINEAR ODE MODELS

concentration curve shows a faster decay rate for higher concentrations of the drug than for lower concentrations. In the simplest case, the rate of metabolism is *linear*, or proportional to the concentration of the drug. The proportionality constant is called the *first-order metabolic rate*.

Let us build an ODE model for a simplified pharmacokinetics situation. Suppose that a drug is administered at a constant rate of M (concentration units per time unit) and that it is metabolized at a rate proportional to its plasma concentration C with metabolic rate constant k. Then the ODE model of the concentration of the drug over time $C(t)$ is

$$\frac{dC}{dt} = M - kC$$

The two rate constants M and k have different dimensions, which you should be able to determine yourself. The ODE can be solved using the separate-and-integrate method:

1. Divide both sides by the right-hand side $M - kC$, and multiply both sides by the differential dt:

$$\frac{dC}{M - kC} = dt$$

2. Integrate both sides with respect to the different variables. Don't forget the integration constant!

$$\int \frac{dC}{M - kC} = \int dt \Rightarrow$$

$$-\frac{1}{k} \log |M - kC| = t + A$$

3. Solve for the dependent variable $C(t)$:

$$\exp(\log |M - kC|) = -\exp(kt + A) \Rightarrow$$

$$M - kC = Be^{-kt} \Rightarrow$$

$$C(t) = \frac{M}{k} - Be^{-kt}$$

Notice that I changed the values of integration constants A and B during the derivation, which shouldn't matter, because they have not been determined yet.

4. Plug in $t = 0$, and use the initial value $x(0)$ to solve for the integration constant. If we know the initial value $C(0) = C_0$, then we can plug it in and get the following algebraic expression:

$$C_0 = \frac{M}{k} - B \Rightarrow$$

$$B = C_0 - \frac{M}{k}$$

Then the complete solution is

$$C(t) = \frac{M}{k} - (C_0 - \frac{M}{k})e^{-kt}$$

The solution predicts that after a long time, the plasma concentration will approach the value M/k, since the exponential term decays to zero. Notice that mathematically this is the same type of solution we obtained in equation 15.4 for a generic linear ODE with a constant term.

Discussion questions:

The following questions encourage you to think critically about the pharmacokinetic model presented in this subsection.

Discussion 15.4.1. Describe in words the dependence of the long-term plasma concentration of the drug on the parameters. Does this prediction make intuitive sense?

Discussion 15.4.2. Explain in practical terms the assumption that the administration of the drug results in a constant rate of growth of the concentration. Under what circumstances does this match reality?

Discussion 15.4.3. Explain in practical terms the assumption that the drug metabolism rate is proportional to plasma concentration. Under what circumstances does this match reality?

Discussion 15.4.4. Discuss how you could modify the ODE model to describe other circumstances, or to add other effects to it.

15.4. APPLICATIONS OF LINEAR ODE MODELS

15.4.2 Cole's membrane potential model

In this example we construct and analyze a model of electric potential across a membrane. The potential is determined by the difference in concentrations of charged particles (ions) on the two sides of the phospholipid bilayer. The ions can flow through specific channels across the membrane, changing the concentration and thus the electric potential. K. S. Cole used principles of electrical circuits to devise the first quantitative model of the membrane voltage (Cole and Cole 1941), which eventually led to more sophisticated models of Hodgkin and Huxley, and others.

To start, we review the physical concepts and laws describing the flow of charged particles. The amount of charge (number of charged particles) is denoted by Q. The rate of flow of charge per time is called the current:

$$I = \frac{dQ}{dt}$$

Current can be analogized to the flow of a liquid, and the difference in height that drives the liquid flow is similar to the electric potential, or voltage. The relationship between voltage and current is given by *Ohm's law:*

$$V = IR$$

where R is the *resistance* of an electrical conductor, and sometimes we use the *conductance* $g = 1/R$ in the relationship between current and voltage:

$$gV = I$$

There are devices known as capacitors, which can store a certain amount of electrical potential in two conducting plates separated by a dielectric (nonconductor). The voltage drop across a capacitor is described by the capacitor law

$$V_C = \frac{Q}{C}$$

where C is the *capacitance*, and Q is the charge of the capacitor.

Lipid bilayer membranes separate media with different concentrations of ions on the two sides, typically the extracellular and cytoplasmic sides. The differences in concentrations of different ions produce a membrane potential. The membrane itself can be thought of as a capacitor, with two charged layers separated by the hydrophobic fatty-acid tails in the middle. In addition, there are ion channels that allow ions to flow from the side with higher concentration to that with lower (these are known as passive channels, as opposed to active pumps that can transport ions against the concentration gradient, which we neglect for now.) These channels conduct ions in one direction up to the *reversal potential* V_R, but reverse direction at higher voltages. The channels are analogous to conducting metal wires and therefore act as resistors with a specified conductance g. Finally, the electrochemical concentrations of ions act as batteries for each species (Na^+, K^+, etc.) The overall electric circuit diagram of this model is shown in Figure 15.3b. Because the different components are connected in parallel, the total current has to equal the sum of the current passing through each element: the capacitor (membrane) and the gated resistors (specific ion channels). The current flowing through a capacitor can be found by differentiating the capacitor law:

$$\frac{dQ}{dt} = I = C\frac{dV_C}{dt}$$

The current flowing through each ionic channel is described by

$$I = g(V - V_R)$$

The total ion flow through the system is described as follows, where i denotes the different ionic species:

$$I_{app} = C\frac{dV}{dt} + \sum_i g(V - V_{Ri}) \tag{15.6}$$

Let us reduce this model to a simple version, where there is no applied current ($I_{app} = 0$) and only a single ionic species with reversal potential V_R. Then the differential equation looks like

$$\frac{dV}{dt} = -\frac{g}{C}(V - V_R) \tag{15.7}$$

15.4. APPLICATIONS OF LINEAR ODE MODELS

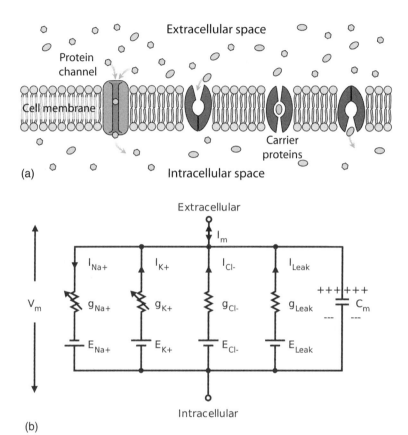

Figure 15.3: (a) A cell membrane with ion channels. Image by LadyofHats in public domain via Wikimedia Commons. (b) Model of a membrane as an electrical circuit with ion channels as resistors and the membrane as a capacitor. Image by NretsNrets under CC BY-SA 3.0 via Wikimedia Commons.

The Cole membrane potential ODE cam also be solved by the separate-and-integrate method. Dividing both sides by $V - V_R$ and multiplying through by the differential dt, we get

$$\frac{dV}{V - V_R} = -\frac{g}{C} dt$$

Remember that V_R is a constant, while $V(t)$ is the dependent variable. Performing the substitution $u = V - V_R$ and integrating, we get $\ln|V - V_R| = -\frac{g}{C}t + A$, where A is the integration constant. Exponentiating both sides and solving for $V(t)$, we obtain

$$V(t) = V_R + Ae^{-\frac{g}{C}t}$$

So, if we began with voltage potential of V_0 at time 0, we have $V_0 = V_R + A \Rightarrow A = V_0 - V_R$. Thus, the complete solution is:

$$V(t) = V_R + (V_0 - V_R)e^{-\frac{g}{C}t} \tag{15.8}$$

This model predicts that if there is no applied current, then starting at a voltage V_0, the membrane potential will exponentially decay (or grow) to the channel's resting (or reversal) potential.

Notice how similar the solutions of this model is to the pharmacokinetic model, even though they are modeling different phenomena. This is an illustration of the power of mathematical modeling, which allows us to use the same tools to draw general conclusions. The biggest takeaway from this observation is that solutions of linear ODEs, whether homogeneous or not, are always exponential in their time dependence. In cases with a constant rate term, the solutions also include a constant term, to which the solution converges if the exponential term has a negative constant in its exponent.

Discussion questions:

The following questions encourage you to think critically about the membrane potential model.

Discussion 15.4.5. The solution of the simplified membrane potential model predicts the convergence of the potential to a constant value. Does this prediction make intuitive sense?

15.5. COMPUTATIONAL PROJECTS

Discussion 15.4.6. Describe what assumptions were made to turn a cell membrane into an electric circuit. Discuss whether they are biologically realistic.

Discussion 15.4.7. To simplify the model, we neglected the applied current term by setting it to zero. Rewrite the ODE with this term added, and predict what effect it will have on the solution.

Discussion 15.4.8. We also simplified the model to contain only one kind of ion channel. Rewrite the ODE to contain two types of ion channels, and solve the new ODE. How is the solution different from the one for a single channel?

15.5 Computational projects

15.5.1 error and time step

The linear population model

$$\frac{dN}{dt} = (b-d)N \tag{15.9}$$

has per capita birth and death rates (b and d, respectively), which may be expressed as a percentage population change per year (e.g., $d = 0.12$ is a death rate of 12% per year). Assume that the initial population is 100,000 individuals, and perform the following tasks.

Tasks

1. Write a Forward Euler script to solve this ODE for any given values of b and d (they should be inputs into your defining function). Set $b = 0.5$ and $d = 0.12$, and report whether the solution behaves as you would expect it to, for $Tmax = 50$ and $dt = 0.1$. Describe the behavior of the solution of the ODE over time.

2. Change the birth rate to $b = 0.01$, and run the calculation again for the same values of $Tmax$ and dt. Compare the behavior of the solution of the ODE to the one in the previous task.

3. Calculate the numerical solution of the ODE with different time steps: $dt = 1, 5, 10$. Report how the solution behaves for larger values of the time step. What time step would you consider optimal for this problem?

4. Find the analytic solution of the ODE for $b = 0.5$ and $d = 0.12$. Use your FE code to plot the numerical solution with $dt = 0.1$ and the analytic solution on the time interval from 0 to 50 years. Calculate and plot the vector of errors (difference between the analytic and numerical solutions), and plot it over time. Describe how the error behaves in this plot.

5. Compute the average error of the numerical solution as the sum of absolute values of the differences between the analytic solution vector x and the numerical solution vector y, divided by the total number of time steps. Keeping the same parameters $b = 0.5$, and $d = 0.12$, calculate the average error of the numerical solution for five values of the time step: $dt = 0.1, 0.05, 0.01, 0.005, 0.001$ and $Tmax = 50$. Report by what factor the error changes for each decrease in the time step.

6. Change the birth rate to $b = 0.01$, and again calculate and plot the numerical solution along with the analytic solution for the same values of $Tmax$ and dt. Calculate and plot the vector of errors (difference between the analytic and numerical solution), and plot it over time. Describe how the error behaves in this plot.

7. Keeping parameters $b = 0.01$ and $d = 0.12$, repeat task 5: calculate the average error of the numerical solution for five values of the time step: $dt = 0.1, 0.05, 0.01, 0.005, 0.001$ and $Tmax = 50$. Report by what factor the error changes for each decrease in the time step.

15.5.2 pharmacokinetics model

The following ODE is a model for the rate of change of concentration of acetaminophen (in $\mu g/mL$), where A is the acetaminophen con-

15.5. COMPUTATIONAL PROJECTS

centration, k is the first-order rate constant of elimination, and $S(t)$ is the (possibly time-dependent) rate of administration of the drug:

$$\frac{dA}{dt} = -kA + S(t)$$

The elimination half-life of acetaminophen is about 3 hours, which corresponds to a first-order rate constant of $k = 0.23$.

Tasks

1. Use your Forward Euler script with the defining function modified to match the equation above with $S(t) = 0$ to find the numerical solution of this model, with initial concentration $A(0) = 15$ over 24 hours and time step $dt = 0.1$. Describe how the solution behaves in the long term, try a couple of different initial values, and comment on whether the solution depends on the initial value. Then gradually increase the time step dt, and report at what value the numeric solution substantially changes (goes from smooth to ragged).

2. Suppose that acetaminophen is administered at the constant rate of 5 μg/mL/hour. Modify the term S(t) in the defining function to reflect this information, and calculate the numerical solution with initial concentration $A(0) = 15$ for 24 hours with timestep $dt = 0.1$. Describe how the solution behaves in the long term, try a couple of different initial values, and comment on whether it depends on the initial value. Again, gradually increase the time step and report at what value the numeric solution substantially changes.

3. Suppose that acetaminophen is administered at the sinusoidal rate of $2 + 2\sin(t)$ μg/mL/hour. Modify the term $S(t)$ in the defining function to reflect this information, and calculate the numerical solution with initial concentration $A(0) = 0$, for 48 hours with $dt = 0.1$. Describe how the solution behaves in the long term, try a couple of different initial values, and comment on whether the solution depends on the initial value. Again, gradually increase the time step, and report at what value the numerical solution substantially changes.

Chapter 16

Graphical analysis of ordinary differential equations

> *I find the great thing in this world is not so much where we stand, as in what direction we are moving.*
> —Oliver Wendell Holmes, Sr., *The Autocrat of the Breakfast Table*

We now proceed from linear ODEs to more complicated nonlinear equations. In contrast to linear differential equations, which can be solved in general, nonlinear differential equations might not be solvable even theoretically. Even though the solutions cannot be written down, they exist and can exhibit much more interesting behaviors than the exponential solutions we have seen. When solutions cannot be found on paper, we have two options: (1) use qualitative or graphical tools, such as finding equilibrium points and their stability, to predict the long-term dynamics of the solution; or (2) construct numerical solutions that approximate the true solution. In this chapter we concentrate on the qualitative approach to analyzing ODEs, which allows one to predict the behavior of solutions of any autonomous ODE based on the graph of the defining function of the equation. In this chapter you will learn to do the following:

1. find equilibrium values of an ODE,

2. analyze the stability of equilibria based on the graph of the defining function,

3. write down stability conditions analytically,

4. use graphical techniques to predict the behavior of the solution of a difference equation without solving it, and

5. understand basic compartment epidemiology models.

16.1 ODEs with nonlinear terms

The simple, linear population growth models we have seen in Chapters 14 and 15 assume that the per capita birth and death rates are constant; that is, they stay the same regardless of population size. The solutions for these models either grow or decay exponentially, but in reality, populations do not grow without bounds. It is generally true that the larger a population grows, the more scarce resources become, and so survival becomes more difficult. For larger populations, this could lead to higher death rates, lower birth rates, or both.

How can we incorporate this effect into a quantitative model? We will assume there are separate birth and death rates, and that the birth rate declines as the population grows, while the death rate increases. Suppose there are inherent birth rates b and d, and the overall birth and death rates B and D depend linearly on population size P: $B = b - aP$ and $D = d + cP$.

To model the rate of change of the population, we need to multiply the rates B and D by the population size P, since each individual can reproduce or die. Also, since the death rate D decreases the population, we need to put a negative sign on it. The resultant model is

$$\dot{P} = BP - DP = [(b-d) - (a+c)P]P$$

The parameters of the model, the constants a, b, c, d, have different meanings. Performing dimensional analysis, we find that b and d

16.2. QUALITATIVE ANALYSIS OF ODES

have the dimensions of $1/[t]$, the same as the rate r in the exponential growth model. However, the dimensions of a (and c) must obey the relation: $[P]/[t] = [a][P]^2$, and thus,

$$[a] = [c] = \frac{1}{[t][P]}$$

This shows that the constants a and c have to be treated differently than b and d. Let us define the inherent growth rate of the population to be $r_0 = b - d$ (if the death rate is greater than the birth rate, the population will inherently decline). Then let us introduce another constant K, such that $(a + c) = r_0/K$. It should be clear from the dimensional analysis that K has units of P, population size. Now we can write down the logistic equation in the canonical form:

$$\dot{P} = r\left(1 - \frac{P}{K}\right)P \tag{16.1}$$

This model can be rewritten as $\dot{P} = aP - bP^2$, so it is clear that there is a *linear term* (aP) and a *nonlinear term* $(-bP^2)$. When P is sufficiently small (and positive) the linear term is greater, and the population grows. When P is large enough, the nonlinear term dominates, and the population declines.

It should be apparent that there are two fixed points, at $P = 0$ and at $P = K$. The first one corresponds to a population with no individuals. In contrast, K signifies the population at which the negative effect of population size balances out the inherent population growth rate; it is called the *carrying capacity* of a population in its environment (Otto and Day 2007). Next we analyze the qualitative behavior of the solution without explicitly writing it out.

16.2 Qualitative analysis of ODEs

In this section we analyze the behavior of solutions of an autonomous ODE without solving it on paper. Generally, ODE models for realistic biological systems are nonlinear, and most nonlinear differential equations cannot be solved analytically. We can make predictions about the behavior, or *dynamics* of solutions by considering the properties of the *defining function*, which is the function on the right-hand side of a general autonomous ODE:

$$\frac{dx}{dt} = f(x)$$

16.2.1 graphical analysis of the defining function

The defining function relates the value of the solution variable x to its rate of change dx/dt. For different values of x, the rate of change of $x(t)$ is different, and it is defined by the function $f(x)$. There are only three options:

- if $f(x) > 0$, $x(t)$ is increasing at that value of x
- if $f(x) < 0$, $x(t)$ is decreasing at that value of x
- if $f(x) = 0$, $x(t)$ is not changing that value of x

To determine for which values of x the solution $x(t)$ increases and decreases, it enough to look at the plot of $f(x)$. On the intervals where the graph of $f(x)$ is above the x-axis, $x(t)$ increases; on the intervals where the graph of $f(x)$ is below the x-axis, $x(t)$ decreases. The roots (zeros) of $f(x)$ are special cases: they separate the range of x into the intervals where the solution grows and where it decreases. This seems exceedingly simple, and it is, but it provides specific information about $x(t)$, without having to know how to write down its formula.

For an autonomous ODE with one dependent variable, the direction of the rate of change prescribed by the differential equation can be graphically represented by sketching the *flow on the line* of the dependent variable. The "flow" stands for the direction of change at every point: specifically, increasing, decreasing, or not changing. The flow is plotted on the horizontal x-axis, so if x is increasing, the flow will be indicated by a rightward arrow, and if it is decreasing, the flow will point to the left. The fixed points separate the regions of increasing (rightward) flow and decreasing (leftward) flow.

16.2. QUALITATIVE ANALYSIS OF ODES

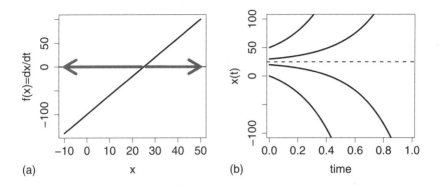

Figure 16.1: (a) Plot of the defining function of the ODE $\dot{x} = 4x - 100$ with direction of flow $x(t)$ indicated with arrows on the x-axis; (b) plot of solutions $x(t)$ of the ODE staring with four initial values.

Example. Consider a linear ODE the like we solved in section 15.2:
$$\frac{dx}{dt} = 4x - 100$$

Its defining function is a straight line vs x. Its graph is shown in Figure 16.1a. Based on this graph, we conclude that the solution decreases when $x < 25$ and increases when $x > 25$. Thus we can sketch the solution $x(t)$ over time, without knowing its functional form. The dynamics depends on the initial value: if $x(0) < 25$, the solution will keep decreasing without bound and go off to negative infinity; if $x(0) > 25$, the solution will keep decreasing without bound and go off to positive infinity. This is shown by plotting numerical solutions of this ODE for several initial values in Figure 16.1b. The dashed line shows the location of the special value of 25, which separates the interval of growth from the interval of decline.

Example. Now let us analyze a nonlinear ODE; specifically, the logistic model with the following parameters:
$$\frac{dP}{dt} = 0.3P\left(1 - \frac{P}{40}\right)$$

The defining function is a downward-facing parabola with two roots at $P = 0$ and $P = 40$, as shown in Figure 16.2a. Between the two roots, the defining function is positive, which means the derivative dP/dt is positive too, so the solution grows on that interval. For $P < 0$ and $P > 40$, the solution decreases. Therefore, we can sketch the graphs of the solution $P(t)$ starting with different initial conditions, as shown in Figure 16.2b.

To summarize, the defining function of the ODE determines the rate of change of the solution $x(t)$ depending on the value of x. The graphical approach to finding areas of right and left flow is based on graphing the function $f(x)$ and dividing the x-axis based on the sign of $f(x)$. In the areas where $f(x) > 0$, its graph is above the x-axis, and the flow is to the right; conversely, when $f(x) < 0$, its graph is below the x-axis, and the flow is to the left. The next subsection puts this approach in a more analytic framework.

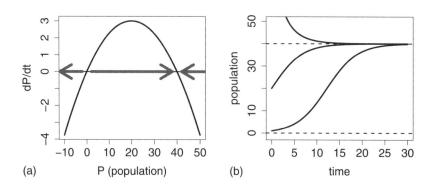

Figure 16.2: (a) Plot of the defining function of the ODE $\dot{P} = 0.3P(1 - P/40)$ with direction of flow of $P(t)$ indicated by arrows on the P-axis; (b) plot of solutions $P(t)$ of the ODE starting with three initial values.

16.2.2 fixed points and stability

We have seen that the dynamics of solutions of differential equations depend on the initial value of the dependent variable: for some

16.2. QUALITATIVE ANALYSIS OF ODES

values the solution increases, for others it decreases, and for intermediate values it remains the same. Those special values separating intervals of increase and decrease are called *fixed points* (or equilibria), and the first step to understanding the dynamics of an ODE is finding its fixed points. A fixed point is a value of the solution at which the dynamical system stays constant; thus, the derivative of the solution must be zero. Here is the formal definition.

Definition 16.1. For a differential equation $\dot{x} = f(x)$, a point x^* that satisfies $f(x^*) = 0$ is called a *fixed point* or *equilibrium*, and the solution with the initial condition $x(0) = x^*$ is constant over time: $x(t) = x^*$.

For instance, the linear equation $\dot{x} = rx$ has a single fixed point at $x^* = 0$. For a more interesting example, consider a logistic equation: $\dot{x} = x - x^2$. Its fixed points are the solutions of $x - x^2 = 0$, therefore it has two fixed points: $x^* = 0, 1$. We know that if the solution has either of the fixed points as the initial condition, it will remain at that value for all time.

Locating the fixed points is not sufficient to predict the global behavior of the dynamical system, however. What happens to the solution of a dynamical system if the initial condition is very close to an equilibrium but not precisely at it? Put another way, what happens if the equilibrium is *perturbed*? The solution may be attracted to the equilibrium value (i.e., it approaches it ever closer), or else it is not. In the first case, this is called a *stable equilibrium*, because a small perturbation does not dramatically change the long-term behavior of the solution. In the latter case, the equilibrium is called unstable, and the solution perturbed from the equilibrium never returns. These concepts are formalized in the following definition (Strogatz 2001).

Definition 16.2. A fixed point x^* of an ODE $\dot{x} = f(x)$ is called a *stable fixed point or sink* if for a sufficiently small number ϵ, the solution $x(t)$ with the initial condition $x_0 = x^* + \epsilon$ approaches the fixed point x^* as $t \to \infty$. If the solution $x(t)$ does not approach x^* for all nonzero ϵ, the fixed point is called *an unstable fixed point or source*.

To determine whether a fixed point is stable analytically, we use an approach called *linearization*, which involves replacing the function $f(x)$ with a linear approximation. Let us define $\epsilon(t)$ to be the deviation of the solution $x(t)$ from the fixed point x^*, so we can write $x(t) = x^* + \epsilon(t)$. Assuming that $\epsilon(t)$ is small, we can write the function $f(x)$ using Taylor's formula:

$$f(x^* + \epsilon(t)) = f(x^*) + f'(x^*)\epsilon(t) + \cdots = f'(x^*)\epsilon(t) + \cdots$$

The term $f(x^*)$ vanishes, because it is zero by definition 16.1 of a fixed point. The ellipsis indicates all the terms of order $\epsilon(t)^2$ and higher, which are very small if $\epsilon(t)$ is small and thus can be neglected. So we can write the following approximation to the ODE $\dot{x} = f(x)$ near a fixed point:

$$\dot{x} = \frac{d(x^* + \epsilon(t))}{dt} = \dot{\epsilon}(t) = f'(x^*)\epsilon(t)$$

Thus we replaced the complicated nonlinear ODE near a fixed point with a linear equation, which approximates the dynamics of the deviation $\epsilon(t)$ near the fixed point x^*; note that the derivative $f'(x^*)$ is a constant for any given fixed point. In section 15.2 we classified the behavior of solutions for the general linear ODE $\dot{x} = rx$, and now we apply this classification to the behavior of the deviation $\epsilon(t)$. If the multiple $f'(x^*)$ is positive, the deviation $\epsilon(t)$ is growing, the solution is diverging away from the fixed point, and thus the fixed point is unstable. If the multiple $f'(x^*)$ is negative, the deviation $\epsilon(t)$ is decaying, the solution is converging to the fixed point, and thus the fixed point is stable. Finally, there is the borderline case of $f'(x^*) = 0$, which is inconclusive, and the fixed point may be either stable or unstable. The derivative stability analysis is summarized as follows:

- $f'(x^*) > 0$: the slope of $f(x)$ at the fixed point is positive; then the fixed point is unstable
- $f'(x^*) < 0$: the slope of $f(x)$ at the fixed point is negative; then the fixed point is stable
- $f'(x^*) = 0$: stability cannot be determined from the derivative

16.2. QUALITATIVE ANALYSIS OF ODES

Therefore, knowing the derivative or the slope of the defining function at the fixed point is enough to know its stability. If the derivative has the discourtesy of being zero, the situation is tricky, because then higher-order terms that we neglected make the difference. We will mostly avoid such borderline cases, but they are important in some applications (Strogatz 2001).

One word of caution: the derivative of the defining function $f'(x)$ is not the second derivative of the solution $x(t)$. This is a common mistake, because the function $f(x)$ is equal to the time derivative of $x(t)$. However, the derivative $f'(x)$ is not with respect to time, it is with respect to x, the dependent variable. In other words, it reflects the slope of the graph of the defining function $f(x)$, not the curvature of the graph of the solution $x(t)$.

To summarize, here is an outline of the steps for analyzing the behavior of solutions of an autonomous one-variable ODE. These tasks can be accomplished either by plotting the defining function $f(x)$ and finding the fixed points and their stability based on the plot, or by solving for the fixed points on paper, then finding the derivative $f'(x)$ and plugging in the values of the fixed points to determine their stability. Either approach is valid, but the analytical methods are necessary when dealing with models that have unknown parameter values, which makes it impossible to represent the defining function in a plot.

Outline of qualitative analysis of an ODE

- Find the fixed points by setting the defining function $f(x) = 0$ and solving for values of x^*.

- Divide the domain of x into intervals separated by fixed points x^*.

- Determine on which interval(s) the solution $x(t)$ is increasing and on which it is decreasing.

- Use derivative stability analysis (graphically or analytically) to determine which fixed points are stable.

- Sketch the solutions $x(t)$ starting at different initial values, based on the stability analysis and whether the solution is increasing or decreasing in a particular interval.

Example: Linear model. Consider the linear ODE that we analyzed above $dx/dt = 4x - 100$. Let us go through the steps of qualitative analysis:

- Find the fixed points by setting the defining function to 0: $0 = 4x - 100$, so there is only one fixed point $x^* = 25$.

- Divide the domain of x into intervals separated by fixed points x^*: the intervals are $x < 25$ and $x > 25$.

- The solution is decreasing on the interval $x < 25$, because $f(x) < 0$ there; the solution is increasing on the interval $x > 25$, because $f(x) > 0$.

- The derivative $f'(x)$ at the fixed point is 4, so the fixed point is unstable.

- Solutions $x(t)$ starting at different initial values are shown in Figure 16.1b, and they behave as follows. Solutions with initial values below $x^* = 25$ are decreasing, and those with initial values above $x^* = 25$ are increasing.

Example: Logistic model. Consider the logistic model from the previous subsection; $dP/dt = 0.3P(1-P/40)$. We have analyzed the stability of the two fixed points using the plot in Figure 16.2 and saw that the flow takes the solution away from $P = 0$ and toward $P = K$. Thus the first fixed point is unstable, while the second is stable. Let us repeat the analysis using analytical tools:

- Find the fixed points by setting the defining function to 0: $0 = 0.3P(1 - P/40)$. The two solutions are $P^* = 0$ and $P^* = 40$.

- Divide the domain of P into intervals separated by fixed points P^*: the intervals are $P < 0$; $0 < P < 40$; and $P > 40$.

- The solution is decreasing on the interval $P < 0$, because $f(P) < 0$ there. The solution is increasing on the interval $0 < P < 40$, because $f(P) > 0$; and the solution is decreasing for $P > 40$, because $f(P) < 0$ there.

16.2. QUALITATIVE ANALYSIS OF ODES

- The derivative is $f'(P) = 0.3 - 0.3P/20$; since $f'(0) = 0.3 > 0$, the fixed point is unstable; since $P'(40) = -0.3 < 0$, that fixed point is stable.

- Solutions $P(t)$ starting at different initial values are shown in Figure 16.2b, and they behave as follows. Solutions with initial values below $P^* = 0$ are decreasing, those with initial values between 0 and 40 are increasing and asymptotically approaching 40, and those with initial values above 40 are decreasing and asymptotically approaching 40.

Example: Semi-stable fixed point. Consider the ODE $dx/dt = -x^3 + x^2$, whose defining function is plotted in Figure 16.3a, showing two fixed points at $x = 0, 1$.

- Find the fixed points by setting the defining function to 0: $0 = -x^3 + x^2$. The two fixed points are $x^* = 0$ and $x^* = 1$.

- Divide the domain of x into intervals separated by fixed points x^*: the intervals are $x < 0$; $0 < x < 1$; and $x > 1$.

- The solution is increasing on the interval $x < 0$, because $f(x) > 0$ there; the solution is increasing on the interval $0 < x < 1$, because $f(x) > 0$; and the solution is decreasing for $x > 1$, because $f(x) < 0$ there.

- The derivative is $f'(x) = -3x^2 + 2x$; since $f'(0) = 0$, the stability of this fixed point is indeterminate; since $f'(1) = -1 < 0$, this fixed point is stable.

- The solutions $x(t)$ starting at different initial values are shown in Figure 16.3b, and they behave as follows. Solutions with initial values below 0 are increasing and asymptotically approaching 0, those with initial values between 0 and 1 are increasing and asymptotically approaching 1, and those with initial values above 1 are decreasing and asymptotically approaching 1.

This example shows how graphical analysis can help when derivative analysis is indeterminate. The arrows on the x-axis of Figure 16.3 show the direction of the flow in the three different regions separated by the fixed points. Flow is to the right for $x < 1$, to the left for for $x > 1$; it is clear that the arrows approach the fixed point from both sides, and thus the fixed point is stable, as the negative slope of $f(x)$ at $x = 1$ indicates. In contrast, the fixed point at $x = 0$ presents a more complicated situation. The slope of $f(x)$ is zero, and the flow is rightward on both sides of the fixed point. This type of fixed point is sometimes called *semi-stable*, because it is stable when approached from one side and unstable when approached from the other.

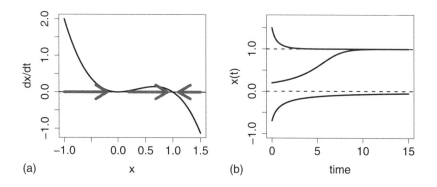

Figure 16.3: (a) Plot of the defining function of the ODE $\dot{x} = -x^3 + x^2$ with direction of flow of $x(t)$ indicated by arrows on the x-axis; (b) plot of solutions $x(t)$ of the ODE starting with three initial values.

Math exercises:

For the following differential equations, (a) plot the defining function over the indicated range (use any computational tools you wish) to determine the intervals on which the dependent variable is increasing and decreasing; (b) find the equilibria; (c) determine the stability of each equilibrium; (d) based on your analysis in parts (a–c), sketch (by hand) plots of the solutions with the specified initial values.

16.2. QUALITATIVE ANALYSIS OF ODES

Exercise 16.2.1.
$$\frac{dN}{dt} = 0.2N - 1200;\ N \in (0, 10000);\ N(0) = 800;\ N(0) = 8000$$

Exercise 16.2.2.
$$\frac{dP}{dt} = 0.3P(40 - P);\ P \in (-1, 60);\ P(0) = 50;\ P(0) = 10$$

Exercise 16.2.3.
$$\frac{dZ}{dt} = Z^2 - 50Z;\ Z \in (-1, 70);\ Z(0) = 3;\ Z(0) = 60$$

Exercise 16.2.4.
$$\frac{dI}{dt} = 0.1I(1 - I) - 0.03I;\ I \in (-0.1, 1.1);\ I(0) = 0.2;\ I(0) = 0.9$$

Exercise 16.2.5.
$$\frac{dR}{dt} = \frac{R}{1+R} - 0.1R;\ R \in (-0.1, 10);\ R(0) = 20;\ R(0) = 0$$

Exercise 16.2.6.
$$\frac{dP}{dt} = 0.02P(P - 100)(1200 - P)\ P \in (-0.1, 1200);\ P(0) = 20;$$
$$P(0) = 1000$$

Exercise 16.2.7.
$$\frac{dY}{dt} = 0.01Y(Y - 100)(Y - 200)\ Y \in (-0.1, 300);\ Y(0) = 20;$$
$$Y(0) = 250$$

Thinking problem:

Refer back to section 2.2 for the algebraic form of the logistic function

Problem 16.2.1. The logistic function was defined in chapter 2, equation 2.4. Show that it is a solution of the logistic ODE defined in equation 16.1 by differentiating the function and substituting it into the equation. How do the parameters a, b, and r correspond to the parameters r and K?

16.3 Modeling infectious disease spread

The field of *epidemiology* studies the distribution of disease and health states in populations. Epidemiologists describe and model these issues with the goal of helping public health workers devise interventions to improve the overall health outcomes on a large scale. One particular topic of interest is the the spread of infectious disease and how best to respond to it. Because epidemiology is concerned with large numbers of people, the models used in the field do not address the details of an individual disease history. One approach to modeling this is to put people into categories, which is called a *compartment model*. We already saw an example of such models in Chapter 10 with a Markov model describing a person with two states: susceptible and infected. Dividing people into categories involves the assumption that everyone in a particular category behaves in the same manner: for instance, all susceptible people are infected at the same rate and all infected people recover at the same rate.

Let us construct an ODE to describe a two-compartment epidemiology model. There are two dependent variables to be tracked: the number of susceptible (S) and infected (I) individuals in the population. The susceptible individuals can get infected, while the infected ones can recover and become susceptible again. The implicit assumption is that there is no immunity, and recovered individuals can become infected with the same ease as those who have never been infected. There are some human diseases for which this is true, for instance, the common cold or gonorrhea. This is called an SIS model of infections disease. Transitions between the different classes of individuals can be summarized by

$$S + I \xrightarrow{\beta} I \xrightarrow{\gamma} S$$

Here β is the individual rate of infection, also known as the transmission rate, and γ is the individual rate of recovery. There is an important distinction between the processes of infection and recovery: the former requires an infected individual and a susceptible individual, while the latter needs only an infected individual. Therefore, it is reasonable to make the assumption that the rate of growth of

16.3. MODELING INFECTIOUS DISEASE SPREAD

infected individuals is the product of the individual transmission rate β and the product of the number of infected and susceptible individuals. The overall rate of recovery is the individual recovery rate γ multiplied by the number those infected. This leads to the following two differential equations:

$$\dot{S} = -\beta IS + \gamma I$$
$$\dot{I} = \beta IS - \gamma I$$

Note that, as in the chemical kinetics models, the two equations add up to zero on the right-hand side, leading to the conclusion that $\dot{S} + \dot{I} = 0$. Therefore, the total number of people is a conserved quantity N, which does not change. This makes sense, since we did not consider any births or deaths in the ODE model, only transitions between susceptible and infected individuals.

We can use the conserved quantity N to reduce the two equations to one, by the substitution of $S = N - I$:

$$\dot{I} = \beta I(N - I) - \gamma I$$

This model may be analyzed using qualitative methods that were developed in this chapter, allowing prediction of the dynamics of the fraction of infected for different transmission and recovery rates. First, let us find the fixed points of the differential equation. Setting the equation to zero, we find

$$0 = \beta I(N - I) - \gamma I \Rightarrow I^* = 0; \ I^* = N - \gamma/\beta$$

This means that there are two equilibrium levels of infection: either nobody is infected ($I^* = 0$), or there is some persistent number of infected individuals ($I^* = N - \gamma/\beta$). Notice that the second fixed point is only biologically relevant if $N > \gamma/\beta$.

Use the derivative test to check for stability. First, find the general expression for derivative of the defining function: $f'(I) = -2\beta I + \beta N - \gamma$.

The stability of the fixed point $I^* = 0$ is found by plugging this value into the derivative formula: $f'(0) = \beta N - \gamma$. We learned in subsection 16.2.2 that a fixed point is stable if the derivative of the defining function is negative at that point. Therefore, $I^* = 0$

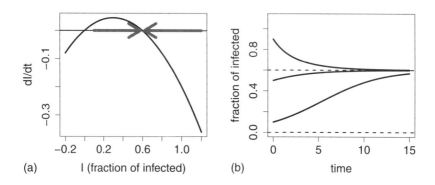

Figure 16.4: Graphical analysis of the SIS model with I representing the fraction of infected individuals ($N = 1$), $\beta = 0.5$, and $\gamma = 0.2$; (a) plot showing the flow of the solutions on the I-axis, with a stable equilibrium at 0.6 and an unstable equilibrium at 0; (b) three solutions of the model starting at three different initial values all converge to the same fraction of infected.

is stable if $\gamma - \beta N > 0$, and is unstable otherwise. This gives us a *stability condition* on the values of the biological parameters. If the recovery rate γ is greater than the rate of infection for the population (the transmission rate multiplied by the population size) βN, then the no-infection equilibrium is stable. This predicts that the infection dies out if the recovery rate is faster than the rate of infection, which makes biological sense.

Similarly, we find the stability of the second fixed point $I^* = N - \gamma/\beta$ by substituting its value into the derivative, to obtain $f'(N - \gamma/\beta) = \gamma - \beta N$. By the same logic as above, this fixed point is stable if $\gamma - \beta N < 0$, or if $\gamma < \beta N$. This is a complementary condition for the fixed point at 0; that is, only one fixed point can be stable for any given parameter values. In the biological interpretation, if the transmission rate βN is greater than the recovery rate γ, then the epidemic will persist.

We can use our graphical analysis skills to illustrate the situation. Consider a situation in which $\gamma < \beta N$. As predicted by

16.3. MODELING INFECTIOUS DISEASE SPREAD

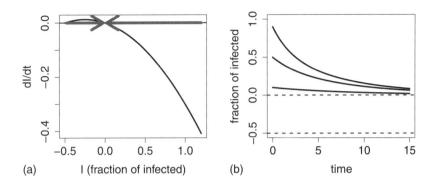

Figure 16.5: Graphical analysis of the SIS model with I representing the fraction of infected individuals ($N = 1$), $\beta = 0.2$, and $\gamma = 0.3$; (a) plot showing the flow of the solutions on the I-axis, with a stable equilibrium at 0 and an unstable equilibrium at -0.5; (b) three solutions of the model starting at three different initial values all converge to 0 infected.

stability analysis, the zero infection equilibrium should be unstable, and the equilibrium at $N - \gamma/\beta$ should be stable. To plot the function $f(I) = \beta I(N - I) - \gamma I$, choose the specific parameter values $N = 1$, $\gamma = 0.1$, and $\beta = 0.2$; setting $N = 1$ means S and I represent the fractions of the population in the susceptible and infected categories, respectively. Figure 16.4a shows the direction of the flow on the I-axis prescribed by the defining function $f(I)$ with arrows. It is clear that solutions approach the fixed point at $N - \gamma/\beta$ from both directions, which make it a stable fixed point, while diverging from $I = 0$, as shown in Figure 16.4b.

In contrast, if $\gamma > \beta N$, stability analysis predicts that the no-infection equilibrium ($I = 0$) is stable. Figure 16.5a shows the plot of the defining function for the parameter values $N = 1$, $\gamma = 0.3$, and $\beta = 0.2$. The flow on the I-axis is toward the zero equilibrium; therefore it is stable. Note that the second equilibrium at $I^* = N - \gamma/\beta$ is negative and thus has no biological significance. The solutions, if the initial value is positive, all approach 0, so the infection inevitably dies out.

Mathematical modeling of epidemiology has been a success story over the past few decades. Public health workers routinely estimate the parameter called the *basic reproductive ratio* R_0, defined to be the average number of new infections caused by a single infected individual in a susceptible population. This number comes from our analysis above, where we found $R_0 = N\beta/\gamma$ to determine whether an epidemic persists (Brauer and Castillo-Chavez 2011). This number is critical in more sophisticated models of epidemiology.

Mathematical models are used to predict the time course of an epidemic, called the *epidemic curve*, and then advise on the public health interventions that can reduce the number of affected individuals. In reality, most epidemic curves have the shape similar to the data from the Ebola virus epidemic shown in Figure 16.6. Most such curves show an initial increase in infections, peaking, and then declining to low levels, which is fundamentally different than the solution curves we obtained from the two-compartment model. To describe dynamics of this nature, models with more than two variables are needed, such as classic three-compartment susceptible-infected-recovered (SIR) models and their modifications (Brauer and Castillo-Chavez 2011). Being able to predict the future of an epidemic based on R_0 and other parameters allows public health officials to prepare and deploy interventions (vaccinations, quarantine, etc.) that have the best shot at minimizing the epidemic.

Discussion questions:

The following questions encourage you to think critically about modeling infectious diseases.

Discussion 16.3.1. What effect does changing the infection rate β have on the basic reproductive rate? Explain the biological intuition behind this.

Discussion 16.3.2. What effect does changing the recovery rate γ have on the basic reproductive rate? Explain the biological intuition behind this.

Discussion 16.3.3. Discuss what assumptions are made by using compartment models and when they might be justified.

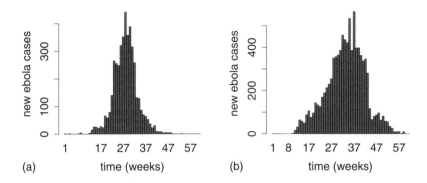

Figure 16.6: Number of new cases of Ebola virus infections per week in (a) Liberia and (b) Sierra Leone. Time ranges from March 17, 2014 (week 1) to May 20, 2015 (week 61). Data from http://apps.who.int/gho/data/node.ebola-sitrep.

Discussion 16.3.4. Discuss the difference in assumptions in using a Markov model with Susceptible and Infected compartments compared to an ODE model with the same two compartments. Under what circumstances does it make sense to use one or the other?

Discussion 16.3.5. Read the paper "Mathematical assessment of the effect of traditional beliefs and customs on the transmission dynamics of the 2014 Ebola outbreaks" (Agusto, Teboh-Ewungkem, and Gumel 2015) and discuss the strengths and limitations of the more complicated compartment model intended to account for human behavior.

16.4 Computational projects

In these projects you will use graphical analysis of ODEs to find fixed points and their stability. This requires plotting the defining function of the ODE, as shown in the following example script. The last line plots a horizontal line $y = 0$ to help locate the fixed points.

```
# Define the function:
DefFunk <- function(x, a, b) {
    ans <- a * x^2 - b
    return(ans)
}
a <- 1
b <- 4
x <- seq(-5, 5, 0.1)    # define the array of values of x
# calculate the values of y using a function
y <- DefFunk(x, a, b)
# plot the function
plot(x, y, t = "l", xlab = "x", ylab = "dx/dt")
abline(0, 0)    # draw a line for y=0
```

Notice that I used two input variables in the defining function (a and b) as parameters that can be changed *outside* the function. This is a good coding practice—the best way to use a function is by changing its inputs, not by monkeying with its code.

After qualitative analysis, you will check whether solutions behave in the predicted way by running Forward Euler to calculate the numerical solution. You can use the code from the assignment in Chapter 15 to compute the numerical solution: simply change the function name inside the for loop.

16.4.1 logistic population growth model

This is the logistic model of population growth, with P representing population size, r is the growth rate parameter, and K is the carrying capacity parameter; P is measured in thousands, and time is measured in years.

$$\frac{dP}{dt} = rP(1 - P/K)$$

Tasks

1. For the values $r = 0.3$ and $K = 40$, plot the graph of the defining function (right-hand side of the ODE) using R over some

16.4. COMPUTATIONAL PROJECTS

interval that includes all zeros of the function. Based on the graph, find the equilibria of the ODE, determine their stability, and predict the behavior of the solution for the following initial values: $P(0) = 1$; $P(0) = 39$; $P(0) = 20$.

2. For the values $r = 0.02$ and $K = 120$, plot the graph of the defining function (right-hand side of the ODE) using R over some interval that includes all zeros of the function. Based on the graph, find the equilibria of the ODE, determine their stability, and predict the behavior of the solution for the following initial values: $P(0) = 10$; $P(0) = 50$; $P(0) = 150$.

3. Write a script that uses the Forward Euler method to solve the logistic ODE for any given values of r and K. Use this script to solve the logistic ODE with parameters $r = 0.3$ and $K = 40$, with a small time step (e.g., $dt = 0.1$) and the following initial values: $P(0) = 1$; $P(0) = 39$; $P(0) = 20$. Plot the numerical solutions of the ODE over a sufficiently large time $Tmax$ to observe convergence. Do the solution plots look consistent with your prediction in task 1?

4. Use your Forward Euler script to solve the logistic ODE with parameters $r = 0.02$ and $K = 120$, with a small time step (e.g., $dt = 0.01$) and the following initial values: $P(0) = 10$; $P(0) = 50$; $P(0) = 150$. Plot the numerical solutions of the ODE over a sufficiently large time $Tmax$ to observe convergence. Do the solution plots look consistent with your prediction in task 2?

5. Let $r = 0.3$ and $K = 40$, and start at $P(0) = 1$. Gradually increase dt until you see the solution behaving differently than expected. Report at what value of dt this strange behavior begins, investigate what happens for even larger time steps, and describe what happens to the solutions.

16.4.2 SIS epidemic model

The following ODE is a simple model of an infectious epidemic with only two kinds of individuals: susceptible and infected; the total population size stays the same. The variable I is the fraction of individuals in the population who are infected, and the parameters β and γ are the infection and recovery rates, respectively; time is measured in days (Britton 2004):

$$\frac{dI}{dt} = \beta I(1-I) - \gamma I$$

Tasks

1. For the values $\beta = 0.1$ and $\gamma = 0.03$, plot the graph of the defining function (right-hand side of the ODE) using R over the interval $[0, 1]$. Based on the graph, find the equilibria of the ODE, determine their stability, and predict the behavior of the solution for the following initial values: $I(0) = 0.01$; $I(0) = 0.8$; $I(0) = 0.5$.

2. For the values $\beta = 0.1$ and $\gamma = 0.5$, plot the graph of the defining function (right-hand side of the ODE) using R over the interval $[0, 1]$. Based on the graph, find the equilibria of the ODE, determine their stability, and predict the behavior of the solution for the following initial values: $I(0) = 0.01$; $I(0) = 0.8$; $I(0) = 0.5$.

3. Use your script for Forward Euler from Chapter 15 to solve the SIS model ODE for any given values of β and γ. Use this script to solve the SIS model ODE with parameters $\beta = 0.1$ and $\gamma = 0.03$, with a small time step (e.g., $dt = 0.1$), and the following initial values: $I(0) = 0.01$; $I(0) = 0.8$; $I(0) = 0.5$. Plot the numerical solutions of the ODE over a sufficiently large time $Tmax$ to observe convergence. In each case, report whether the solution approaches zero (the epidemic burns out), how the behavior depends on the initial value, and comment on whether the solution dynamics agrees with your prediction in task 1.

4. Use your Forward Euler script to solve the SIS model ODE with parameters $\beta = 0.1$ and $\gamma = 0.5$, with a small time step (e.g., $dt = 0.1$) and the following initial values: $I(0) = 0.01$;

$I(0) = 0.8$; $I(0) = 0.5$. Plot the numerical solutions of the ODE over a sufficiently large time $Tmax$ to observe convergence. In each case, report whether the solution approaches zero (the epidemic burns out), how the behavior depends on the initial value, and comment on whether the solution dynamics agrees with your prediction in task 2.

5. For $I(0) = 0.5$ and $\beta = 0.1$; starting with $\gamma = 0.3$, progressively decrease the value of the recovery rate γ, and report the behavior of the numerical solution (with a small enough dt and a large enough $Tmax$). Report the critical value of γ at which the solution converges to a positive number in the long run (the epidemic becomes self-sustaining). Keep decreasing the recovery rate, and report what happens to the equilibrium fraction of infected people. Does it ever get to 1?

Chapter 17

Chaos and bifurcations in difference equations

Everything was in confusion in the Oblonskys' house.
—Leo Tolstoy, *Anna Karenina*

In this chapter we will analyze nonlinear discrete dynamical systems. Their solutions, as those of nonlinear ODEs, exhibit much more interesting behaviors than the exponential solutions of linear equations and are typically not solvable analytically. There may be multiple fixed points, some stable and others unstable, and even crazier behaviors are possible that are not permitted in smooth-flowing ODEs. Specifically, we will see solutions that oscillate, and those that behave without any pattern at all, which are called chaotic. You will learn to do the following in this chapter:

1. find equilibrium values of nonlinear discrete-time models,

2. analyze the stability of equilibria based on the graph of the updating function,

3. write down stability conditions analytically,

4. use graphical techniques to predict the behavior of the solution of a difference equation without solving it, and

5. understand what the term "chaos" means.

17.1 Logistic model in discrete time

Here we analyze the famous discrete-time logistic model. Its assumptions are similar to those used to derive the logistic ODE. We arbitrarily choose a simple nonconstant relationship between the birth and death rates and the size of the population:

$$b = b_1 - b_2 N(t); \; d = d_1 + d_2 N(t)$$

We model both the birth and death rates as linear functions of population size, but with the birth rate declining and the death rate increasing with larger population size. We can now substitute these birth and death rates into the difference equation for population size:

$$N(t+1) - N(t) = (b-d)N(t) = [(b_1 - d_1) - (b_2 + d_2)N(t)]N(t)$$

A simpler way of writing this equation is to let $r = 1 + b_1 - d_1$ and $K = b_2 + d_2$, leading to the following iterated map:

$$N(t+1) = (r - KN(t))N(t) \qquad (17.1)$$

This is discrete time version the *logistic model* of population growth. As in the ODE model see, it has two different parameters, r and K. If $K = 0$, the equation reduces to the old linear population model. Intuitively, K is the parameter describing the effect of increasing population on the population growth rate. Let us analyze the dimensions of the two parameters by writing down the dimensions of the variables of the difference equation. The dimensional equation is

$$N(t+1) = [population] = [r - KN(t)]N(t) =$$
$$= ([r] - [K] \times [population]) \times [population]$$

Matching the dimensions on the two sides of the equation leads us to conclude that the dimensions of r and k are different:

$$[r] = 1; \; [K] = \frac{1}{[population]}$$

The difference equation for the logistic model is *nonlinear*, because it includes a second power of the dependent variable. In general, it is difficult to solve nonlinear equations, but we can still say a lot about this model's behavior without knowing its explicit solution.

17.2 Qualitative analysis of difference equations

17.2.1 fixed points or equilibria

We have seen that the solutions of difference equations depend on the initial value of the dependent variable. For linear difference equations, the long-term behavior of the solution does not depend dramatically on the initial condition. Just as for ODEs, there are special values of the dependent variable for which the dynamical system is constant.

Definition 17.1. For a difference equation $X(t+1) = f(X(t))$, a point X^* which satisfies $f(X^*) = X^*$ is called a *fixed point* or *equilibrium*. If the initial condition is a fixed point, $X_0 = X^*$, the solution will stay at the same value for all time, $X(t) = X^*$.

The reason these special points are also known as equilibria is due to the precise balance between growth and decay that is mandated at a fixed point. In terms of population modeling, at an equilibrium, the birth rates and death rates are equal. Speaking analytically, to find the fixed points of a difference equation, one must solve the equation $f(X^*) = X^*$. It may have none, or one, or many solutions.

Example. The linear population models that we analyzed in Chapter 14 have the mathematical form $N(t+1) = rN(t)$ (where r can be any real number). Then the only fixed point of those models is $N^* = 0$, that is, a population with no individuals. If there are any individuals present, we know that the population will grow to infinity if $|r| > 1$, and decay to 0 if $|r| < 1$. This is true even for the smallest population size, as long as it is not exactly zero.

Example. Let us go back to the example of a linear difference equation with a constant term, which we solved numerically in section 14.4. The equation is $N(t+1) = -0.5N(t) + 15$, and we saw that the numerical solutions all converged to the same value, regardless of the initial value. Let us find the equilibrium value of this model using the definition

$$N^* = -0.5N^* + 15 \Rightarrow 1.5N^* = 15 \Rightarrow N^* = 15/1.5 = 10$$

If the initial value is equal to the equilibrium, $N(0) = 10$, then the solution will remain constant for all time, since the next value $N(t+1) = -0.5 \times 10 + 10 = 15$ remains the same. This is the value to which the solutions converge in the numerical solutions we saw in the example in section 14.4.

Example: Discrete logistic model. Let us use the simplified version of the logistic equation $N(t+1) = r(1 - N(t))N(t)$ and set the right-hand side function equal to the variable N to find the fixed points N^*:

$$r(1 - N^*)N^* = N^*$$

There are two solutions to this equation, $N^* = 0$ and $N^* = (r-1)/r$. These are the fixed points (or equilibrium population sizes) for the model, the first being the obvious case when the population is extinct. The second equilibrium is more interesting, as it describes the *carrying capacity* of a population in a particular environment. If the initial value is equal to either of the two fixed points, the solution will remain at that same value for all time. But what happens to solutions that do not start a fixed point? Do they converge to a fixed point, and if so, to which one?

17.2.2 stability of fixed points

What happens to the solution of a dynamical system if the initial condition is very close to an equilibrium but not precisely at it? Put another way, what happens if the equilibrium is *perturbed*? To answer the question, we no longer confine ourselves to the integers, to be interpreted as population sizes. We will instead consider, abstractly, what happens if the smallest perturbation is added to a fixed point. Will the solution tend to return to the fixed point or tend to move away from it? The answer to this question is formalized in the following definition (Strogatz 2001).

Definition 17.2. For a difference equation $X(t+1) = f(X(t))$, a fixed point X^* is *stable* if for a sufficiently small number ϵ, the solution $X(t)$ with the initial condition $X_0 = X^* + \epsilon$ approaches the fixed point X^* as $t \to \infty$. If the solution $X(t)$ does not approach X^* for any nonzero ϵ, the fixed point is called *unstable*.

17.2. QUALITATIVE ANALYSIS OF DIFFERENCE ...

This definition is identical to the concept of stability of fixed points of ODEs you saw in definition 16.2. The practical implications for solutions are very similar, but with a subtle wrinkle for difference equations. Whereas ODE solutions cannot reach a fixed point in finite time, for a difference equation there may be certain initial conditions, distinct from the fixed points, such that the solution ends up at the equilibrium value after a finite number of time steps. It is even possible for a solution to end up at an unstable fixed point, having started elsewhere! This is a consequence of the essential jumpiness of the solutions of difference equations.

As you saw in section 16.2, stability can be determined using *linearization*. Take a general difference equation written in terms of some function $X(t+1) = f(X(t))$. Let us define the *deviation* from the fixed point X^* at time t to be $\epsilon(t) = X(t) - X^*$. Then we can use the linear (first-order) Taylor approximation at the fixed point and write down the following expression:

$$X(t+1) = f(X^*) + \epsilon(t)f'(X^*) + \cdots$$

The ellipsis means that the expression is approximate, with terms of order $\epsilon(t)^2$ and higher swept under the rug. Since we take $\epsilon(t)$ to be small, those terms are very small and can be neglected. Since X^* is a fixed point, $f(X^*) = X^*$. Thus, we can write the following difference equation to describe the behavior of the deviation from the fixed point X^*:

$$X(t+1) - X^* = \epsilon(t+1) = \epsilon(t)f'(X^*)$$

Thus we started out with a general function defining the difference equation and transformed it into a linear equation for the deviation $\epsilon(t)$. Note that the multiplicative constant here is the derivative of the function at the fixed point: $f'(X^*)$.

We found the general solution for linear difference equations in section 14.2, which we can use to describe the behavior of the perturbation to the fixed point. The behavior depends on the value of the derivative of the updating function $f'(X^*)$:

- $|f'(X^*)| > 1$: the deviation $\epsilon(t)$ grows, and the solution moves away from the fixed point; the fixed point is *unstable*

- $|f'(X^*)| < 1$: the deviation $\epsilon(t)$ decays, and the solution approaches the fixed point; the fixed point is *stable*

- $|f'(X^*)| = 1$: the fixed point may be stable or unstable, and more information is needed

The classification of the behavior near a fixed point is directly analogous to the analysis of ODEs in section 16.2, except that for ODEs, the stability depends on the sign of the derivative, but in discrete time it depends on whether its absolute value is greater than or less than 1.

Example: Linear difference equations. Let us analyze the stability of the fixed point of the linear difference equation $N(t+1) = -0.5N(t)+15$. The derivative of the updating function is equal to -0.5. Because it is less than 1 in absolute value, the fixed point is stable, so solutions converge to this equilibrium, as we saw in section 14.2. We can state more generally that any linear difference equation of the form $N(t+1) = aN(t)+b$ has one fixed point, which is equal to $N^* = b/(1-a)$. This fixed point is stable if $|a| < 1$ and unstable if $|a| > 1$.

Example: Discrete logistic model. In subsection 17.2.1 we found the fixed points of the simplified logistic model. To determine what happens to the solution, we need to determine the stability of both equilibria. Since the stability of fixed points is determined by the derivative of the defining function at the fixed points, we compute the derivative of $f(N) = rN - rN^2$ to be $f'(N) = r - 2rN$, and evaluate it at the two fixed points $N^* = 0$ and $N^* = (r-1)/r$:

$$f'(0) = r; \quad f'((r-1)/r) = r - 2(r-1) = 2 - r$$

Because the intrinsic death rate cannot be greater than the birth rate, we know that $r > 0$. Therefore, we have the following stability conditions for the two fixed points:

- the fixed point $N^* = 0$ is stable for $r < 1$ and unstable for $r > 1$;

- the fixed point $N^* = (r-1)/r$ is stable for $1 < r < 3$ and unstable otherwise.

17.3 Graphical analysis

Many ecological models are far more sophisticated than the logistic model (Otto and Day 2007). However, a surprising amount of complex behavior is hidden in this simple quadratic equation. We turn to computational means to further investigate it.

17.3.1 graphical analysis using R

Let us use the computational power of R to visualize the dynamics of solutions of the discrete logistic model. The code below defines the updating function with two input arguments: the variable N and the parameter r. This is an important programming practice: all variables and parameters that are used in a function should be passed to it through the function call. It is a bad idea to assign any variables inside the function, unless they never change, because then one has to modify the function itself to change the variable. Another bad idea is to use outside variables in the function without passing them. This will (usually) work in R, but it is sloppy and can lead to problems if the function is copied and moved to some other script where the outside variable has not been previously defined. The following script calls the updating function to compute the solution of the logistic model, after setting the parameter r and the initial value of the solution vector `sol`.

```
logistic.funk <- function(N, r) {
    ans <- r * (1 - N) * N
    return(ans)
}
numsteps <- 20
r <- 1.5   # set growth parameter
sol <- rep(0, numsteps + 1)
sol[1] <- 0.5   # initialize solution
time <- 1:(numsteps + 1)
for (i in 1:numsteps) {
    sol[i + 1] <- logistic.funk(sol[i], r)
}
```

The solution is now stored in the vector sol, which we can plot over time. The script below also plots the updating function along with the identity line. The two graphs are shown in Figure 17.1 for the value of $r = 1.5$. The graph of the definition function (Figure 17.1a) shows two fixed points at $N = 0$ and $N = 0.33$. The first one is unstable, because the slope of the updating function is greater than 1, and the second one stable, because the slope of the updating function is less than 1—this is clear in comparison to the graph of the identity line which has slope 1. Figure 17.1b shows the solution over time converging to the stable fixed point at $N = 0.33$. In the next subsection we investigate the solutions using a graphical approach and observe the effect of changing the parameter r on the dynamics.

17.3.2 cobweb plots

In addition to calculating numerical solutions, computational tools can be used to analyze dynamical systems graphically. A lot of information can be gleaned by plotting the graph of the defining function of an iterated map $X(t+1) = f(X(t))$, as we saw in section 17.2. Here is a summary of what we can learn from the graph of the function $f(x)$:

- The location of the fixed points of the iterated map. Since the condition for a fixed point is $f(x) = x$, they can be found at the intersections of the graph of $y = f(x)$ and $y = x$ (the identity straight line).

- The stability of fixed points. The derivative of $f(x)$ at a fixed point determines its stability. Graphically, this means that the slope of $f(x)$ at the point of intersection with $y = x$ can be used for this purpose; if it is steeper (in absolute value) than the straight line $y = x$, then the fixed point is unstable, but if its slope is less than one in absolute value, the equilibrium is stable.

- Graphical iteration of the difference equation. The value of the function $f(x)$ gives the value of x at the next time step,

17.3. GRAPHICAL ANALYSIS

```
x <- seq(0, 1, 0.01)
y <- logistic.funk(x, r)
# plot the updating function
plot(x, y, xlab = "N(t)", ylab = "f(N) = r*(1-N)*N",
    t = "l", lwd = 3, cex = 1.5, cex.axis = 1.5,
    cex.lab = 1.5)
lines(x, x, col = 2, lwd = 3, lty = 2)
# plot the numeric solution over time
plot(time, sol, t = "b", xlab = "time",
    ylab = "N(t) (population)",
    lwd = 3, cex = 1.5, cex.axis = 1.5, cex.lab = 1.5)
```

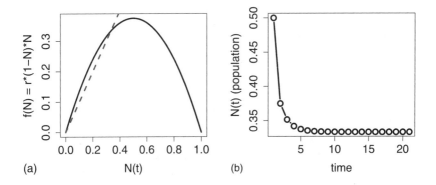

Figure 17.1: Plots of (a) the updating function and (b) the solution of the logistic model $N(t+1) = rN(t)(1-N(t))$ with $r = 1.5$.

and this fact can be used to produce a graph of successive values of the dependent variable: $x_0, x_1, x_2, ...$

Let us exploit the idea in the last point for graphical analysis of an iterated map. Starting with some initial condition X_0, the value of x_1 is given by $f(X_0)$. To show this graphically, starting at the point X_0 on the x-axis, draw a vertical line to $y = f(X_0)$. Next, draw a horizontal line to the graph of $y = x$. Since the y-and x-coordinates are equal, we now have the value of $x_1 = f(X_0)$ as the x-coordinate. Then, repeat the process by drawing a vertical line to $y = f(x_1)$ and a horizontal line $y = x$, and so forth. This results in what is known as a *cobweb plot*, which can be summarized as follows.

1. Plot the graph of the defining function $y = f(x)$.
2. Graph the identity line $y = x$ on the same plot.
3. Start with an initial value of the variable x_0.
4. Draw a vertical line to graph of the defining function $f(x_0)$.
5. Draw a horizontal line to the graph of the line $y = x$, and set the new value x_1.
6. Draw a vertical line to the defining function $f(x_1)$.
7. Draw a horizontal line to the graph of the line $y = x$, and set the new value x_2.
8. Repeat as long as needed.

The resulting sequence of x-coordinates is a numerical solution of the difference equation and provides a visual means of assessing its dynamics. For instance, the values may converge to a fixed point, grow to infinity, or bounce around without settling down. We now have at our disposal analytical, numerical, and graphical tools to analyze and predict the behavior of a dynamical system.

Example. The following script implements the cobweb plot algorithm using the `lines()` command to draw alternating horizontal and vertical lines. I define auxiliary variables `current.y` and `next.y` to designate the beginning and the end of each vertical segment. At the beginning of each iteration, `next.y` is set to the value of the updating function at the current variable value, and at the end of each iteration, `current.y` becomes `next.y`.

```
x <- seq(0, 1, 0.01)    # range of the plot
r <- 1.5   # set the parameter
plot(x, logistic.funk(x, r), type = "l", xlab = "N(t)",
    ylab = "f(N) = r*(1-N)*N", lwd = 3, lty = 1)
```

17.3. GRAPHICAL ANALYSIS

```
lines(x, x, col = 2, lwd = 3, lty = 2)
numsteps <- 10    # number of iterations
sol <- rep(0, numsteps + 1)   # initialize solution vector
sol[1] <- 0.1    # set initial value
current.y <- 0   # set current y-coordinate
for (i in 1:numsteps) {
    sol[i + 1] <- logistic.funk(sol[i], r)   # next value
    next.y <- sol[i + 1]   # next y-coordinate
    lines(x = c(sol[i], sol[i]), y = c(current.y, next.y))
    lines(x = c(sol[i], next.y), y = c(next.y, next.y))
    current.y <- next.y
}
```

Figure 17.2 shows the cobweb plots on two different logistic updating functions, one with parameter $r = 0.8$, and the other with parameter $r = 1.8$. In both plots, we see the cobweb stair-stepping briskly toward a single stable fixed point. In the case of $r = 0.8$, the stable equilibrium value is 0, and in the case of $r = 1.8$, the stable equilibrium is around 0.33. This is what we expected both from the stability analysis in the last section and from the numerical solutions above.

Now let us increase the parameter value and see how it affects the cobweb plot. Figure 17.3 shows cobweb plots for the logistic difference model with $r = 2.8$ and $r = 3.3$. In Figure 17.3a, the cobweb plot approaches the stable fixed point around 0.6, but the cobweb plot is not approaching the fixed point in a straightforward manner. Instead, the solution oscillates around it, converging toward it eventually. In Figure 17.3b, the fixed point has lost its stability, and solution does not converge to any single point, instead making a box-looking plot. This corresponds to the solution oscillating between two values (since the box has only two x-coordinates) for as long as the simulation is run. This is called a *period-two oscillation*.

Math exercises:

For the following biological scenarios: (a) write down a difference equation model in the updating function form $X_{t+1} = f(X_t)$;

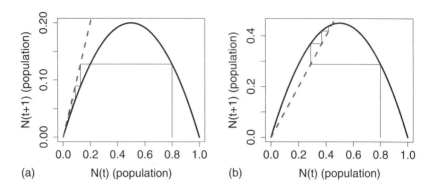

Figure 17.2: Cobweb plots of the logistic model $N(t+1) = r(1 - N(t))N(t)$ with initial value $N(0) = 0.8$ for two different parameter values: (a) $r = 0.8$ and (b) $r = 1.8$.

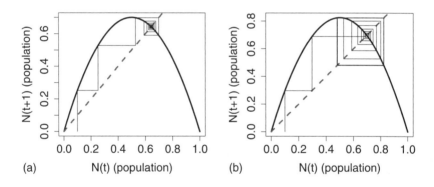

Figure 17.3: Cobweb plots of the logistic model $N(t+1) = r(1 - N(t))N(t)$ with initial value $N(0) = 0.1$ for two different parameter values: (a) $r = 2.8$ and (b) $r = 3.3$.

(b) sketch the graph of the updating function along with the identity line, indicate the fixed points at the intersections of the two graphs, and sketch a cobweb plot to determine the stability of each fixed point (you may use computational tools to aid your sketching);

(c) find the fixed points analytically, use derivative analysis to determine the stability of each fixed point, and comment on whether your stability analysis agrees with the results of the cobweb sketch (it should!).

Exercise 17.3.1. Bacteria divide in 2 every hour, but half of the existing population dies after reproduction.

Exercise 17.3.2. Each pair of rabbits produces 4 offspring every year and the adults have a 0.5 annual death rate (after reproduction), while a python that lives nearby eats exactly 1 rabbit a month.

Exercise 17.3.3. Five fish are added to an aquarium every month, while 90% of those present survive every month, and there is no reproduction.

Exercise 17.3.4. Microorganisms live in an environment with a reproductive rate $b = 32 - 0.1N$, where b is the birth rate (per individual per day), and N is the current number of microorganisms; all of them live for one day and then die.

Exercise 17.3.5. Insects reproduce every year at a rate of 100 offspring per individual per year, but they also get eaten with the predation (death) rate parameter of $P = 500N/(4 + N^2)$ (per individual per year), where N is the current population. (This one is challenging.)

17.4 Discrete-time logistic model and chaos

In this chapter you have learned to analyze the dynamics of solutions of nonlinear discrete-time dynamical systems without solving them on paper. In sections 17.2 and 17.3 we focused on the logistic difference equation as a simple nonlinear model with a rich array of dynamic behaviors. In this section we will summarize the analysis and draw conclusions for difference equation models in biology. These behaviors was brought to the attention of biologists by John Maynard Smith (Smith 1968) and Robert May (May and Oster 1976).

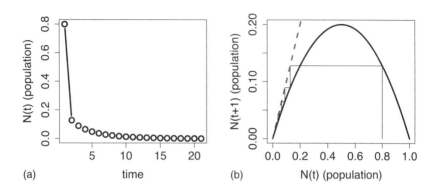

Figure 17.4: (a) Solution of the logistic model $N(t+1) = rN(t)(1-N(t))$ with $r = 0.8$, starting with $N(0) = 0.8$; (b) the updating function with the cobweb plot for the same initial value.

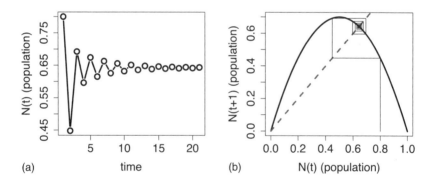

Figure 17.5: (a) Solution of the logistic model $N(t+1) = rN(t)(1-N(t))$ with $r = 2.8$, starting with $N(0) = 0.8$; (b) the updating function with the cobweb plot for the same initial value.

The simplified logistic difference equation $N(t+1) = r(1-N(t))N(t)$ has two fixed points: $N^* = 0$ and $N^* = (r-1)/r$. Stability of these two fixed points depends on the value of the parameter r, which represents the growth rate of the population when

17.4. DISCRETE-TIME LOGISTIC MODEL AND CHAOS

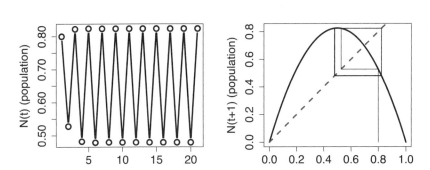

Figure 17.6: (a) Solution of the logistic model $N(t+1) = rN(t)(1 - N(t))$ with $r = 3.3$ starting with $N(0) = 0.8$; (b) the updating function with the cobweb plot for the same initial value.

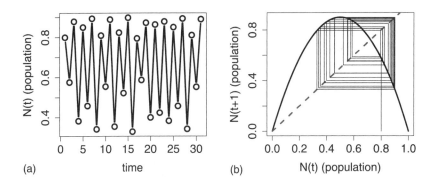

Figure 17.7: (a) A solution of the logistic model $N(t + 1) = rN(t)(1 - N(t))$ with $r = 3.6$ starting with $N(0) = 0.8$; (b) the updating function with the cobweb plot for the same initial value.

the population is low. We found the stability criteria for each of the fixed points in section 17.2, which suggests that we can divide the range of possible values of r (which must be positive to be biological plausible) into three intervals.

Case 1: $r < 1$. The fixed point at $N^* = 0$ is stable, and the fixed point $N^* = (r-1)/r$ is unstable. The solution tends to 0, or extinction, regardless of the initial condition, which is illustrated in Figure 17.4 for $r = 0.8$.

Case 2: $1 < r < 3$. The extinction fixed point $N^* = 0$ is unstable, but the carrying capacity fixed point $N^* = (r-1)/r$ is stable. We can conclude that the solution will approach the carrying capacity for most initial conditions. This is shown in Figure 17.1 for $r = 1.5$ and is illustrated in Figure 17.5 for $r = 2.8$. Notice that although the solution approaches the carrying capacity equilibrium in both cases, when $r > 2$, the solution oscillates while converging to its asymptotic value, foreshadowing the behavior when $r > 3$.

Case 3: $r > 3$. Strange things happen: no stable fixed points exist, so there is no value for the solution to approach. As we saw in section 17.3, the solution can undergo so-called period two oscillations, which are shown in Figure 17.6 with $r = 3.3$. However, even stranger behavior is observed when the parameter r crosses the threshold of about 3.59. Figure 17.7 shows the behavior of the solution for $r = 3.6$, which is no longer periodic, and instead seems to bounce around without any discernible pattern. This dynamics is known as *chaos*.

The last plot of the solution and the cobweb plot show an entirely new and surprising kind of dynamics: not convergence to an equilibrium, not oscillations between several values, not even unbounded growth. Although it has been dubbed chaos, this behavior is not actually random or unpredictable, despite the everyday connotations of the word. If you run the simulation multiple times, starting with the same initial value, the solution will behave exactly the same, since each new value of $N(t+1)$ is computed deterministically from the previous value $N(t)$. However, if the initial value is changed even slightly, the solution looks very different after a few time steps. This characteristic, called *sensitive dependence on initial conditions*, is the telltale quality of chaos. The other necessary ingredient of chaos is its lack of periodicity, which is apparent in the solution in Figure 17.7: although it bounces between similar values, in fact it never repeats the same number exactly.

We can use the results of numerical simulations to plot the long-term solutions for the dependent variable for a range of parameter values, say, between $2.5 < r < 4$. Then we plot the values to which the simulation converged (whether it is one, two, or many) on the y-axis, and the value of the parameter r on the x-axis. The resulting *bifurcation diagram* in shown in Figure 17.8. The value of the parameter r is plotted on the horizontal axis, and the set of values that the dependent variable takes in the long run is shown on the vertical axis. There is only one stable fixed point for $r < 3$, then we see a period-two cycle appear for $3 < r < 3.45$. For values of r greater than about 3.45, a series of period-doubling bifurcations occur with shorter and shorter intervals of r. This is called a *period-doubling cascade*, which culminates at the value of $r \approx 3.57$, where the number of points in the cycle becomes essentially infinite.

What is especially surprising about chaos is that for a given initial condition, a chaotic model gives a completely predictable and reproducible sequence of values of the dependent variable. However, given finite machine precision, or any error in initial conditions, a chaotic system is practically unpredictable and irreproducible. But there is a fundamental difference between deterministic chaos and a stochastic system (e.g., the model of coin tosses where knowing the previous result of the coin flip does not allow us to predict the next result, even under ideal conditions).

Chaos was a popular topic back in the 1980s and 1990s, and it even inspired popular books (Gleick 1988). It is in fact remarkable that very simple difference equations can have solutions of apparently great complexity. This is intriguing, because it appeals to a fairly universal human desire for simple explanations for complicated phenomena. The popular exposure to what was dubbed "chaos theory" (which is not an actual mathematical topic) spawned some inaccurate clichés, such as "a butterfly flapping its wings in South America can cause a hurricane to form and hit Florida." The image refers to the phenomenon of sensitive dependence on initial conditions, but of course it is utterly ridiculous to draw a causal arrow between a butterfly (one of an innumerable number of things that may affect the "initial conditions") and large-scale atmospheric phenomena. Although weather patterns are complex systems that

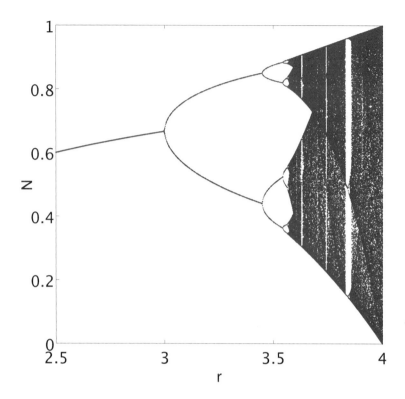

Figure 17.8: Bifurcation diagram for the logistic map $N(t+1) = r(1 - N(t))N(t)$, with the parameter r on the horizontal axis and the vertical axis showing the values of the stable fixed point (for $r < 3$), then the values of the period two oscillation, the period four oscillation, and so forth. For r greater than the critical value the vertical axis shows some of the values the solution chaotically jumps through.

exhibit complex behavior with sensitive dependence on initial conditions, we lack the ability to isolate and control all influences that may perturb it, so pinning it on a butterfly is highly unconvincing.

Despite the initial flurry of excitement, so-called chaos theory has failed to make a big impact on our understanding of complex biological systems. Although it is still quite fascinating intellectually, a simple model like the logistic model is not adequate for

17.4. DISCRETE-TIME LOGISTIC MODEL AND CHAOS

any realistic population, particularly for large values of r where the chaotic behavior occurs. We now appreciate that the essential complexity of biological system requires multiple interacting variables that cannot be reduced to a single equation. However, there has been some successful observation of chaotic behavior in a population of flour beetles, which seemed to agree with predictions of a three-variable difference equation model (Costantino et al. 1997).

We have seen how graphical tools can be used to analyze and predict the behavior of a discrete-time dynamical system. We investigated the logistic model by finding the fixed points and analyzing their stability. Together with analyzing the graph of the updating function and making a cobweb plot, this allows us to describe the dynamics of population growth in the logistic model without doing any "math." The combination of analytical and graphical analyses provides powerful tools for biological modelers.

Discussion questions:

Chaos is difficult to conceptualize, so the following questions aim to help you think through the complexities.

Discussion 17.4.1. Formulate the difference between Markov models and chaotic dynamical systems. What can one predict for a Markov model? What is predictable for a chaotic dynamical model?

Discussion 17.4.2. Qualitatively speaking, why do you think increasing r in the logistic difference equation leads to more wild solutions? Speculate and discuss with your peers.

Discussion 17.4.3. The logistic ODE does not exhibit chaotic behavior for any parameter values. What is the is distinction between the difference and the differential equations that explains this?

Discussion 17.4.4. What biological systems might exhibit chaotic behavior? Speculate which conditions could give rise to the two hallmarks of chaos: sensitive dependence on initial conditions and lack of repetition.

17.5 Computational projects

In these projects you will analyze nonlinear difference equations, which in general can be written as

$$X(t+1) = f((X(t))$$

17.5.1 graphical stability analysis

The graph of the updating function $f(X)$ provides information about the dynamics of the difference equation when plotted together with the identity line $Y = X$. The intersections of those two graphs occur at the equilibrium values or fixed points, and the slope of $f(X)$ at the intersection indicates the stability of the equilibrium.

Tasks

For the following difference equations, do the following: (a) make a plot of the updating function of the difference equation along with the identity line over a range that includes the zeros of the updating function and its intersections with the identity line; (b) identify the equilibrium values based on the plot (the values can be approximate); (c) predict the future dynamics of the solution, starting with the given initial values; (d) compare your prediction with the numerical solutions obtained in the computational projects in Chapter 16.

1. Initial values: $X(0) = 20$; $X(0) = 0$.

$$X(t+1) = 45 - 0.8X(t)$$

2. Initial values: $X(0) = 3$; $X(0) = 1$.

$$X(t+1) = \frac{2}{X(t)+1}$$

3. Initial values: $X(0) = 2$; $X(0) = 9$.

$$X(t+1) = 0.2X(t)(10 - X(t))$$

17.5. COMPUTATIONAL PROJECTS

4. Initial values: $X(0) = 2$; $X(0) = 9$.
$$X(t+1) = 0.4X(t)(10 - X(t))$$

5. Initial values: $X(0) = 2$; $X(0) = 9$.
$$X(t+1) = X(t)(10 - X(t))$$

17.5.2 investigation of chaotic dynamics

We saw in the model $X(t+1) = rX(t)(1-X(t))$ that solutions have qualitatively different dynamics for different values of r. When r is larger than 3, the solutions oscillate between two values, and further increase in r results in different oscillations, and for r above a certain threshold the dynamics are chaotic. In the following tasks you will investigate in detail the onset of chaos in modified logistic difference models. Here is the general form of the model:

$$X(t+1) = rX(t)(K - X(t))$$

Tasks

1. Let $K = 10$ (as in two of the models above). Starting with a small value, increase r and report at what threshold value the dynamics change from converging to one stable fixed point to period two oscillations; call this value r_2.

2. Keeping $K = 10$, further increase r and report at what threshold value the dynamics change from period two to period four oscillations; call this value r_4.

3. Keeping $K = 10$, further increase r and report at what threshold value the dynamics change from period four to period eight oscillations (this will require very small increments in r); call this value r_8.

4. Keeping $K = 10$, further increase r and report at what threshold value the dynamics change from period eight to period sixteen oscillations (this will require very small increments in r and printing out the values of the solution to see whether it converges to eight or sixteen different values); call this value r_{16}.

5. These threshold values $r_2, r_4, r_8, r_{16}, \ldots$ are called the period-doubling cascade, because they keep getting closer to one another. Report the ratios of the successive terms: $r_4/r_2, r_8/r_4, r_{16}/r_8$. Do you see a pattern? Can you predict what the next value r_{32} is going to be?

6. Keeping $K = 10$, further increase r and report at what threshold the dynamics change to chaotic (this will require very small increments in r and printing out the values of the solution to see whether it repeats values after many time steps). Predict the threshold value for the onset of chaos from the period-doubling cascade, and compare the prediction with the results of your numerical experimentation.

Bibliography

Agusto, F. B., M. I. Teboh-Ewungkem, and A. B. Gumel. 2015. "Mathematical assessment of the effect of traditional beliefs and customs on the transmission dynamics of the 2014 Ebola outbreaks." *BMC Medicine* 13: 96–101.

Akiyama, Shuji, Atsushi Nohara, Kazuki Ito, and Yuichiro Maéda. 2008. "Assembly and Disassembly Dynamics of the Cyanobacterial Periodosome." *Molecular Cell* 29 (6): 703–716.

Allman, Elizabeth S., and John A. Rhodes. 2003. *Mathematical Models in Biology: An Introduction.* First edition. Cambridge: Cambridge University Press.

Bodine, Erin N., Suzanne Lenhart, and Louis J. Gross. 2014. *Mathematics for the Life Sciences.* Princeton, NJ: Princeton University Press.

Brauer, Fred, and Carlos Castillo-Chavez. 2011. *Mathematical Models in Population Biology and Epidemiology.* Second edition, 2012. New York: Springer.

Britton, Nicholas F. 2004. *Essential Mathematical Biology.* First edition. New York: Springer.

Center for Constitutional Rights. 2014. "NYPD's Stop and Frisk Practice: Unfair and Unjust." `https://ccrjustice.org/racial-disparity-nypd-stops-and-frisks`.

Chimpanzee Sequencing and Analysis Consortium 2005. "Initial Sequence of the Chimpanzee Genome and Comparison with the Human Genome." *Nature* 437 (7055): 69–87.

Cohen, Joel E. 2004. "Mathematics Is Biology's Next Microscope, Only Better; Biology Is Mathematics' Next Physics, Only Better." *PLoS Biology* 2 (12): e439.

Cole, Kenneth S., and Robert H. Cole. 1941. "Dispersion and Absorption in Dielectrics I. Alternating Current Characteristics." *Journal of Chemical Physics* 9 (4): 341–351.

Costantino, R. F., R. A. Desharnais, J. M. Cushing, and B. Dennis. 1997. "Chaotic Dynamics in an Insect Population." *Science* 275 (5298): 389–391.

Eddy, Sean R. 2004. "What Is a Hidden Markov Model?" *Nature Biotechnology* 22 (10): 1315–1316.

Edelstein-Keshet, Leah. 2005. *Mathematical Models in Biology*. First edition. Philadelphia: Society for Industrial and Applied Mathematics.

Feller, William. 1968. *An Introduction to Probability Theory and Its Applications, Vol. 1*. Third edition. New York and London: Wiley.

Futuyma, Douglas. 2009. *Evolution*. Second edition. Sunderland, MA: Sinauer Associates.

Gleick, James. 1988. *Chaos: Making a New Science*. First edition. London: Penguin.

Goodman, Steven N. 1999 "Toward Evidence-Based Medical Statistics. 1: The P Value Fallacy." *Annals of Internal Medicine* 130 (12): 995–1004.

Green, J. A., P. J. Butler, A. J. Woakes, I. L. Boyd, and R. L. Holder. 2001. "Heart Rate and Rate of Oxygen Consumption of Exercising Macaroni Penguins." *Journal of Experimental Biology* 204 (4): 673–684.

Hayes, B. 2012. "First Links in the Markov Chain." *American Scientist* 101 (2): 92.

BIBLIOGRAPHY

Hou, Yubo, and Senjie Lin. 2009. "Distinct Gene Number–Genome Size Relationships for Eukaryotes and Non-eukaryotes: Gene Content Estimation for Dinoflagellate Genomes." *PLoS ONE* 4 (9): e6978.

Ioannidis, John P. A. 2005a. "Contradicted and Initially Stronger Effects in Highly Cited Clinical Research." *JAMA* 294 (2): 218–228.

———. 2005b. "Why Most Published Research Findings Are False." *PLoS Medicine* 2 (8): e124.

Jain, A., J. Marshall, A. Buikema, T. Bancroft, J. P. Kelly, and C. J. Newschaffer. 2015. "Autism Occurrence by MMR Vaccine Status among Children with Older Siblings with and without Autism." *JAMA* 313 (15): 1534–1540.

Jukes, T. H. and C. R. Cantor. 1969. "Evolution of Protein Molecules." In *Mammalian Protein Metabolism*, edited by M. N. Munro, vol. III, pp. 21–132. New York: Academic Press.

Jungck, John R., Holly Gaff, and Anton E. Weisstein. 2010. "Mathematical Manipulative Models: In Defense of 'Beanbag Biology.'" *CBE Life Sciences Education* 9 (3): 201–211.

Kolokotrones, Tom, Van Savage, Eric J. Deeds, and Walter Fontana. 2010. "Curvature in Metabolic Scaling." *Nature* 464 (7289): 753–756.

Kong, Augustine, Michael L. Frigge, Gisli Masson, Soren Besenbacher, Patrick Sulem, Gisli Magnusson, Sigurjon A. Gudjonsson, et al. 2012. "Rate of de novo Mutations and the Importance of Father's Age to Disease Risk." *Nature* 488 (7412): 471–475.

Leisegang, Rory, Gary Maartens, Michael Hislop, John Sargent, Ernest Darkoh, and Susan Cleary. 2013. "A Novel Markov Model Projecting Costs and Outcomes of Providing Antiretroviral Therapy to Public Patients in Private Practices versus Public Clinics in South Africa." *PLoS ONE* 8 (2): e53570.

Malone, Fergal D., Jacob A. Canick, Robert H. Ball, David A. Nyberg, Christine H. Comstock, Radek Bukowski, Richard L. Berkowitz, et al. 2005. "First-Trimester or Second-Trimester Screening, or Both, for Down's Syndrome." *New England Journal of Medicine* 353 (19): 2001–2011.

Markov, A. A. 1906. "An Extension of the Law of Large Numbers to Quantities Which Depend on Each Other." (Russian) *Bulletein of the Physical-Mathematics Society of Kazan University* 15 (2): 135–156.

———. 1913. "An Example of a Statistical Investigation of the Text of *Eugene Onegin*, Illustrating the Linkage of Trials in a Chain." (Russian) *Bulletin of the Imperial Academy of Sciences* 7 (3): 153–162.

May, Robert M., and George F. Oster. 1976. "Bifurcations and Dynamic Complexity in Simple EcologicalModels." *American Naturalist* 110 (974): 573–599.

Michaelis, L., and M. L. Menten. 1913. "Die Kinetik der Invertinwirkung." *Biochem Z* 49: 333–369.

Miller, Anthony B., Claus Wall, Cornelia J. Baines, Ping Sun, Teresa To, and Steven A. Narod. 2014. "Twenty Five Year Follow-up for Breast Cancer Incidence and Mortality of the Canadian National Breast Screening Study: Randomised Screening Trial." *BMJ* 348: g366.

New York Times. 2014. "On the Stop-and-Frisk Decision: Floyd v. City of New York." http://www.nytimes.com/interactive/2013/08/12/nyregion/stop-and-frisk-decision.html.

Otto, Sarah P., and Troy Day. 2007. *A Biologist's Guide to Mathe- matical Modeling in Ecology and Evolution*. Princeton, NJ: Princeton University Press.

Pevsner, Jonathan. 2009. *Bioinformatics and Functional Genomics*. Second edition. Hoboken, NJ: Wiley-Blackwell.

Podar, Mircea, Iain Anderson, Kira S. Makarova, James G. Elkins, Natalia Ivanova, Mark A. Wall, Athanasios Lykidis, et al. 2008. "A Genomic Analysis of the Archaeal System *Ignicoccus hospitalis–Nanoarchaeum equitans*." *Genome Biology* 9 (11): R158.

Press, William H., Saul A. Teukolsky, William T. Vetterling, and Brian P. Flannery. 2007. *Numerical Recipes: The Art of Scientific Computing*. Cambridge: Cambridge University Press.

Pushkin, Aleksandr. 1832. *Eugene Onegin*.

———. 2009. *Eugene Onegin*. Translated by James Falen. Oxford: Oxford University Press.

Rosenbaum, Sara E. 2011. *Basic Pharmacokinetics and Pharmacodynamics: An Integrated Textbook and Computer Simulations*. First edition. Hoboken, NJ: Wiley.

Schultz, Stanley G. 2002. "William Harvey and the Circulation of the Blood: The Birth of a Scientific Revolution and Modern Physiology." *Physiology* 17 (5): 175–180.

Senn, Stephen. 2011. "Francis Galton and Regression to the Mean." *Significance* 8 (3): 124–126.

Sloan, R. E., and W. R. Keatinge. 1973. "Cooling Rates of Young People Swimming in Cold Water." *Journal of Applied Physiology* 35 (3): 371–375.

Smith, J. Maynard. 1968. *Mathematical Ideas in Biology*. London: Cambridge University Press.

Sonnenberg, Frank A., and J. Robert Beck. 1993. "Markov Models in Medical Decision Making: A Practical Guide." *Medical Decision Making* 13 (4): 322–338.

Strang, Gilbert. 2005. *Linear Algebra and Its Applications*. Fourth edition. Belmont, CA: Brooks Cole.

Strogatz, Steven H. 2001. *Nonlinear Dynamics and Chaos: With Applications to Physics, Biology, Chemistry, and Engineering.* First edition. Boulder, CO: Westview Press.

Thomson, Bob. 2013. "Jukes Cantor Model of DNA Substitution." Workshop in Applied Phylogenetics. http://treethinkers.org/jukes-cantor-model-of-dna-sub stitution/.

Ventura, Mario, Claudia R. Catacchio, Saba Sajjadian, Laura Vives, Peter H. Sudmant, Tomas Marques-Bonet, Tina A. Graves, Richard K. Wilson, and Evan E. Eichler. 2012. "The Evolution of African Great Ape Subtelomeric Heterochromatin and the Fusion of Human Chromosome 2." *Genome Research* 22 (6): 1036–1049.

Watkins, Joseph C. n.d. *An Introduction to the Science of Statistics: From Theory to Implementation.* http://math.arizona.edu/~jwatkins/statbook.pdf

West, Geoffrey B., and James H. Brown. 2005. "The Origin of Allometric Scaling Laws in Biology from Genomes to Ecosystems: Towards a Quantitative Unifying Theory of Biological Structure and Organization." *Journal of Experimental Biology* 208 (9): 1575–1592.

Whitlock, Michael C., and Dolph Schluter. 2008. *The Analysis of Biological Data.* First edition. Greenwood Village, CO: Roberts and Company Publishers.

Index

allele, 17, 59, 92
allometry, 205
amino acids, 58

bacteria
    division, 309, 326
    size, 66
Bayes' formula, 158
Bayesian
    interpretation, 163, 164
    prior probability, 156
Bernoulli trial, 77
bifurcation diagram, 399
binding cooperativity, 38
bioinformatics, 266
blood circulation, 11

cell cycle, 214
central limit theorem, 107, 110
chaos, 398, 402
chi-squared
    assumptions, 144
    distribution, 142
    statistic, 142
confidence interval
    definition, 111
    for mean, 113
    relative risk, 115
contingency table, 126
correlation, 183

covariance, 182
    random variables, 90

data
    categorical, 60, 125
    error bars, 111
    heart rate, 70, 195
    mean, 61
    mutations, 194
    numerical, 60
    range, 63
    standard deviation, 64
    variance, 63
difference equation
    analytic solution, 312
    cobweb plot, 392
    definition, 309
    fixed point, 385
    linearization, 387
    logistic, 384
    nonlinear, 384
    numeric solution, 312
    numerical solution, 319
    solution, 312
    stability, 386
    stability conditions, 389
    updating function, 316, 385
differential equation
    analytic solution, 332
    autonomous, 332

differential equation (cont.)
  defining function, 331, 362
  finite difference methods, 343
  first-order method, 344
  fixed point, 365
  flow, 362
  Forward Euler method, 338
  homogeneous, 332
  linearization, 366
  logistic, 361
  numerical solution, 338
  order, 331
  separate-and-integrate, 333
  SIS model, 371
  stability analysis, 366
  stability condition, 373
dimension, 30
  matrix, 237
dimensional analysis, 30, 311, 329, 360
dimensionless quantity, 31
distribution
  binomial, 90
  chi-squared, 142
  Markov chain, 236
  normal, 107
  uniform, 86
DNA, 57, 295

epidemic
  basic reproductive ratio, 375
  compartment model, 372
  curve, 376
error
  numerical solutions, 343
  roundoff, 343
  truncation, 343

evolution
  chromosome, 299
  genome divergence, 299
  LUCA, 277
  molecular, 294
  natural selection, 276
  phylogenetic distance, 297
  phylogenetic trees, 277
exponential
  function, 35, 286, 334
  growth, 309
  rate constant, 35

fitting, 10
  generalized linear, 203
  goodness, 180
  least squares, 180
  log transform, 200, 208
  over-fitting, 204
  polynomial, 202
Frobenius theorem, 286
function
  call, 317
  exponential, 35, 199, 309, 313
  graph, 34
  in R, 41, 317
  linear, 34
  logistic, 38, 368
  mathematical, 5, 33
  power law, 199, 205
  rational, 37

Galen, 11
genome, 57, 267
  chimpanzee, 298
  exons, 267
  human, 298
  introns, 267

INDEX 413

pseudogenes, 267
sequence, 293

Harvey, William, 13
heart rate, 71
Hill equation, 38
histogram
  definition, 65
  in R, 69
hypothesis
  null, 136
  testing, 136

independence
  Bernoulli trials, 88
  events, 131
  Markov property, 218
  product rule, 134, 220
  random variables, 134
  sampling, 105, 110
initial condition, 312
ion channel, 213
  membrane potential, 353
  nAChR, 217, 230, 251, 271, 302

kinetics
  drug administration, 347
  drug metabolism, 347
  first-order, 52
  law of mass action, 330
  Michaelis-Menten, 52
  pharmacology, 346
  rate constant, 52, 330
  reaction, 51
  zeroth-order, 51

law of large numbers, 105, 106
linear
  difference equation, 312
  differential equation, 333
  function, 34
  slope, 35
linear regression
  assumptions, 181
  residual, 180, 186
  slope, 182
  y-intercept, 183
linearization
  difference equation, 387
  differential equation, 366
logarithmic transform, 199
logistic model, 40
  continuous time, 378
  discrete time, 384

Markov
  absorbing state, 261
  aperiodic state, 261
  cell cycle, 234
  chain, 235
  communicating states, 260
  distribution vector, 235
  history, 254
  HMM, 268
  ion channel model, 217, 230, 251, 271, 302
  property, 218
  recurrent state, 260
  SI model, 229, 250, 270, 301
  stationary distribution, 257
  transient state, 260
  transition diagram, 214

Markov (cont.)
    transition matrix, 219
    transition probability, 216
matrix
    characteristic equation, 282
    definition, 237
    determinant, 278
    diagonal elements, 238
    dimension, 237
    distance, 298
    eigenvalue, 279
    eigenvector, 279
    elements, 237
    multiplication, 238
    second-largest eigenvalue, 287
    trace, 278
mean
    data, 61
    in R, 69
    random variable, 82
    sample, 104
    true, 104
median, 60
    in R, 69
medical
    cancer screening, 162
    heart rate, 67
    maternal age, 71, 126, 146, 194
    paternal age, 71, 194
    prevalence, 158
    relative risk, 115
    testing, 159
    treatment model, 226
membrane potential, 351
    reversal, 352

mode, 60
model
    assumptions, 9
    binding cooperativity, 40
    blood circulation, 15
    cartoon, 10
    cell cycle, 234
    continuous time, 326
    deterministic, 75
    discrete time, 308
    dynamical system, 307
    empirical, 9
    epidemiology, 229, 250, 270, 301, 371
    first-principles, 9
    Hill, 38
    Jukes-Cantor, 249, 293
    logistic, 361, 368
    Malthusian growth, 315
    Markov, 213
    mathematical, 1
    membrane potential, 350
    Michaelis-Menten, 40
    pharmacokinetics, 356
    population, 310, 355
    stochastic, 75
molecular evolution, 249
mRNA, 58
mutation, 92
    de novo, 71, 194
    definition, 58
    missense, 59
    point, 59, 299
    polymorphism, 59
    rate, 248
    substitution, 58, 293

INDEX

nonlinear
    data fitting, 197
    difference equation, 383
    differential equation, 361
nucleotides, 58

parameter, 5, 16, 105
    basic reproductive ratio, 375
    drug metabolism, 350
    fitting, 209
    linear regression, 188
    logistic model, 397
period two oscillation, 393
period-doubling, 399, 402
plot
    in R, 47
    log-log, 200
    residual, 186
    scatter, 67, 178
    semi-log, 199
    time, 67
policy
    MMR vaccination, 115
    stop-and-frisk, 147
population
    bacteria, 326
    birth rate, 310, 355
    carrying capacity, 361, 386
    death rate, 311, 355
    linear model, 310
    logistic model, 378
probability
    axioms, 78
    conditional, 128
    definition, 78
    distribution, 81
    event, 76

prior, 156
sample space, 76
proteins, 58
Pushkin, A.S., 255

R programming
    arithmetic, 20
    chi-squared test, 144
    conditional statement, 166
    data frame, 69
    descriptive statistics, 69
    floating point, 343
    for loop, 98, 117, 225, 246, 320, 342
    function, 41, 317
    function call, 318
    index of vector, 43
    input arguments, 317
    logical test, 166
    machine precision, 22
    matrix diagonalization, 288
    normalizing eigenvectors, 289
    plotting, 47
    scientific notation, 21
    uniform distribution, 116
    variable assignment, 23
    vector variable, 42
random
    computer simulation, 224
    experiment, 76
    number generator, 95, 117, 224
random variable
    binomial, 88
    continuous, 107, 116
    definition, 81

random variable (cont.)
    expected value, 82
    Markov chain, 236
    normal, 107
    standard deviation, 85
    uniform, 86
    variance, 84
randomized controlled trial, 114
regression to the mean, 191
rescaling, 31

sampling
    distribution, 105
    error, 105
scientific research reproducibility, 164
sequencing
    genetic, 2
    next-generation, 2
    whole genome, 193, 299
sets
    complement, 78
    intersection, 78
    union, 78
sickle-cell disease, 59
slope
    linear function, 35
    linear regression, 182
stability
    fixed point, 365, 386
    Forward Euler, 345
standard deviation
    in R, 69
    random variable, 85
standard error, 107
state
    discrete, 213

space, 17
    variable, 17
statistics
    descriptive, 5, 59
    population, 104
    sample, 103
    significance, 143
    simple random sample, 104
systems biology, 2

test
    chi-squared, 143
    false negative, 137
    false positive, 137
    negative predictive value, 159
    negative result, 137
    p-value, 141, 164
    positive predictive value, 159
    positive result, 137
    sensitivity, 138
    specificity, 138
    true negative, 137
    true positive, 137
    type I error, 138
    type II error, 138

units, 30

variable
    assignment, 23
    definition, 16
    dependent, 34, 48, 308, 331
    explanatory, 178
    independent, 34, 48, 308, 331
    response, 178

INDEX

variance
    data, 63
    in R, 69
    random variable, 84
vector
    column, 239
    in R, 42

Markov distribution, 236
row, 239

y-intercept
    linear function, 35
    linear regression, 183